"Always be on the lookout
for the presence of wonder."

E. B. White

JON WATERMAN

Into the Thaw

Witnessing Wonder amid the Arctic Climate Crisis

patagonia®

Into the Thaw
Witnessing Wonder amid the Arctic Climate Crisis

Patagonia publishes a select list of titles on wilderness, wildlife, and outdoor sports that inspire and restore a connection to the natural world and encourage action to combat climate chaos.

Every effort was made to recognize the Indigenous people displaced from their lands by advancing societies. Please e-mail books@patagonia.com with additions or corrections.

Note: Iñupiat, Gwich'in Nàhn, Dënéndeh, Kuuvuan KaNianiq, Inuvialuit, or Inuit Nunangat ancestral lands are denoted in photo captions. If unmentioned, ancestral lands are identified in the preceding photograph caption. While the information is derived from https://native-land.ca, Indigenous people ranged over large areas, and there would have been much overlap in land usage.

Hardcover Edition

Published by Patagonia Works

Printed in Canada on Sustana Enviro 100 Satin FSC certified 100 percent post-consumer paper.

Hardcover ISBN 978-1-952338-23-6
E-Book ISBN 978-1-952338-24-3
Library of Congress Control Number 2024941811

Editor – John Dutton
Photo Editor – Heidi Volpe
Art Director/Designer – Christina Speed
Maps – Christina Speed
Project Manager – Sonia Moore
Photo Production – Bernardo Salce
Production – Natausha Greenblott
Creative Director – Michael Leon
Publisher – Karla Olson

ENDSHEET: Down the Noatak after a six-day trek over the Brooks Range in 2022. A thermokarst landslide on the mountain to the right in the photo caused by permafrost thaw looms above. Gates of the Arctic National Park, Alaska. Dënéndeh, Kuuvuan KaNianiq, Gwich'in Nàhn, and Iñupiat ancestral lands. CHRIS KORBULIC

SECOND SPREAD: An Inuit man praises his *qimmiq* (Eskimo husky) on the sea ice in Elu Inlet in 1999. The fierce qimmiq—more wolf than husky—served for 4,500 years of travel across the Arctic but is now threatened with extinction by snowmachines. JON WATERMAN

TITLE PAGE: Mike Freeman at Icy Reef along the Beaufort Sea coast; the icepack lies just offshore in August 2006, Arctic National Wildlife Refuge. From 1979 to 2023, Arctic summer sea ice has declined by an average of twenty-eight thousand square miles per year. JON WATERMAN

FOLLOWING SPREAD: In the Noatak National Preserve, Chris Korbulic checks out spruce trees collapsing in the permafrost thaw—this "drunken forest" effect is now occurring throughout the North. JON WATERMAN

ENVIRONMENTAL BENEFITS STATEMENT

Patagonia Inc saved the following resources by printing the pages of this book on chlorine free paper made with 100% post-consumer waste.

TREES	WATER	ENERGY	SOLID WASTE	GREENHOUSE GASES
364 FULLY GROWN	29,000 GALLONS	153 MILLION BTUs	1,200 POUNDS	157,000 POUNDS

Environmental impact estimates were made using the Environmental Paper Network Paper Calculator 4.0. For more information visit www.papercalculator.org

1%
FOR THE PLANET.
MEMBER

FSC
www.fsc.org
MIX
Paper | Supporting responsible forestry
FSC® C016245

Also by the Author

Atlas of Wild America
National Geographic Books, 2023

Atlas of the National Parks
National Geographic Books, 2019

Chasing Denali: The Sourdoughs, Cheechakos, and Frauds Behind the Most Unbelievable Feat in Mountaineering
Lyons Press, 2018

Northern Exposures: An Adventuring Career in Stories and Images
University of Alaska Press, 2013

The Colorado River: Flowing Through Conflict
Westcliffe Publishers, 2010

Running Dry: A Journey from Source to Sea Down the Colorado River
National Geographic Books, 2010

Where Mountains Are Nameless: Passion and Politics in the Arctic National Wildlife Refuge
W. W. Norton & Company, 2005

Arctic Crossing: A Journey Through the Northwest Passage and Inuit Culture
Alfred A. Knopf, 2001

The Quotable Climber: Literary, Humorous, Inspirational, and Fearful Moments of Climbing
Lyons Press, 1998

A Most Hostile Mountain: Re-Creating the Duke of Abruzzi's Historic Expedition on Alaska's Mount St. Elias
Henry Holt & Company, 1997

Kayaking the Vermilion Sea: Eight Hundred Miles Down the Baja
Simon & Schuster, 1995

In the Shadow of Denali: Life and Death on Alaska's Mt. McKinley
Doubleday, Bantam, Dell, 1993

Cloud Dancers: Portraits of North American Mountaineers
AAC Press, 1993

High Alaska: A Historical Guide to Denali, Mount Foraker & Mount Hunter
AAC Press, 1989

Surviving Denali: A Study of Accidents on Mount McKinley
AAC Press, 1983

For my sons, Nicholas and Alistair

Caribou antlers on a pass in the Schwatka Mountains above the Noatak headwaters. Every year throughout northwest Alaska, over a hundred thousand caribou-shed antlers—rich with calcium and phosphorus—are gnawed to dust by rodents and other animals seeking the vital minerals. Dënéndeh, Kuuvuan KaNianiq, and Iñupiat ancestral lands. CHRIS KORBULIC

Prologue

A Certain Type of Fun, July 10–12, 2022

Kalulutok Creek would be called a river in most parts of the world. Here in Gates of the Arctic National Park and Preserve, amid the largest span of legislated wilderness in the United States, it's just a creek compared to the massive Noatak River that we're bound for. But in my mind—while we splash-walked packrafts and forded its depths at least thirty times yesterday—Kalulutok will always be an ice-cold, wild river.

It drains the Endicott and Schwatka Mountains, which are filled with the most spectacular granite and limestone spires of the entire Brooks Range. One valley to the east of us is sky-lined with sharp, flinty peaks called the Arrigetch, or "fingers of the outstretched hand" in Iñupiaq.

As the continent's most northerly mountains, the sea-fossil-filled Brooks Range—with more than a half dozen time-worn peaks over eight thousand feet high—is seen on a map as the last curl of the Rocky Mountains before they stairstep into foothills and coastal plains along the Arctic Ocean. The Brooks Range stretches two hundred miles south to north and seven hundred miles to the east, where it jabs into Canada. Although there are more than four hundred named peaks, since the Brooks Range is remote and relatively untraveled, it's rare that anyone bothers to climb these mountains. My river-slogger companion, Chris Korbulic, and I will be one of the summer's handful of exceptions.

PREVIOUS SPREAD: Our cold, 3,900-foot camp beneath an unnamed peak above a lake in the headwaters of Kalulutok Creek. The Brooks Range is filled with hundreds of similarly idyllic and seldom-visited alpine basins. CHRIS KORBULIC

We carry a water filter, but it would be silly to use it. We're higher and farther north than giardiasis-infected beavers and there is no sign of the pellet-dropper deer known as caribou. The creek is fed from the pure ice of shrunken glaciers above and ancient permafrost in the ground below. In what seems like prodigious heat for the Arctic, the taps here are all wide open.

When we get thirsty, we luxuriate in one of the freedoms of a journey through pristine wilds: we kneel, cup our hands, and drink directly from the ice-cream-headache water of the Kalulutok. Its root name, from the word for fish, *kaluk*, has been lost to most Iñupiat, who now mostly speak English. Yet there's no question that the stream is full of grayling, or *sulupaugak*, that flit about as sun-blinkered shadows through the eddies.

If I stand still under a twenty-degree angle from the horizon, the grayling won't see me. A grayling's eyes are remarkably similar to mine—except they can simultaneously focus on near and faraway objects like a hawk—but the optics and underwater reflections limit their vision. I love their eyes: the rim of sparkly gold iris that surrounds the black, teardrop-shaped pupils. When you snatch the fish from their watery world, they stare back with such wide-eyed purity that you wouldn't want to dismiss the presence of their soul.

I also admire their finely sculpted, streamlined form and iridescent blue scales. I repeatedly study these creatures that I share the water with as they restively hold the current in the same way that I stand against a strong wind. They erect their distinctive, sail-shaped dorsal fins and anchor themselves in place against the flow. Then they can be seen as vividly as a worm in mezcal.

Usually less than a foot long and found in northern waters, the grayling is a keystone species, a veritable canary in the coal mine in an era when the Arctic now warms nearly four times faster than the rest of the world. In their migrations up and down tens of thousands

Beneath multiple thermokarst landslides caused by permafrost thaw, we tow our packrafts up Kalulutok Creek to avoid bushwhacking in a brush-thickened valley created by the Greening of the Arctic. CHRIS KORBULIC

of small streams, they provide nutrients throughout the ecosystem and are often the only fish—lithe and muscular—that can wriggle up countless narrow drainages that won't fit the stouter salmon and trout.

While that's the beta from the appreciative fish biologists, they also wonder if this supremely adapted cold-water species will go belly up as the world overheats, streams in the Arctic dry, and summer seasons elongate.[1]

A couple of months from now, as this river-creek begins to crackle with ice buildup, hundreds of *sulupaugak* will migrate downstream into the four hundred-foot-deep waters of Walker Lake. They'll congregate there until spring, preyed upon by the tubby, bully lake trout—akin to grizzlies amid ground squirrels.

I watch the grayling here in the headwaters of Walker Lake and the Kobuk River as I study birds on the wing and other wildlife signs: the wide bear trails bulldozed through the alder thickets, moose scat remarkably similar to Milk Duds, and, in the willows, the ptarmigan excreta like coarse hamburger curls extruded from the meat grinder. On the river bars, braided-rope-shaped wolf feces are also strewn hither and yon.

"TMI, Jon," many people would say. Too Much Information, to stop and talk about and pull apart every piece of animal fecal matter as if it were Play-Doh. Or to badger the Fish and Wildlife Service worker back in Fairbanks on the misconceptions about Arctic char and Dolly Varden. "Char only exist in Alaskan lakes," I insisted, while Chris quietly exited the federal offices, chagrined at my compulsion to debate such arcane details.

My disclaimer: Thirty-nine years ago, I decided to learn all I could about life above the Arctic Circle. As a climber, I traded my worship of high mountains for the High Arctic. I substituted bears and mosquitoes for crevasses and avalanches, but more importantly—like the study of

1 On the North Slope of the Brooks Range, at Alaska's Toolik Field Station, researchers continue a long-term study of how warmed stream temperatures will affect the grayling. Although the jury is still out, a 2006 paper "Climate Change Effects on Hydroecology of Arctic Freshwater Ecosystems" states that earlier breakup and increased air temperatures, along with thaw of the permafrost, will likely warm lake and stream waters, increase aquatic plant production, and release other nutrients that will, in turn, increase organic matter detrimental to Arctic grayling.

crevasse extrication and avalanche avoidance—you couldn't just read about the Arctic or sign up for classroom courses. You have to go on immersive journeys and figure out how the interlocked parts of the natural world fit together. Without guides or someone to hold your hand. Best to hit the ragged edge of exhaustion and make mistakes so that you learn what's important. And to go alone at least once. Along this path, acts of curiosity out on the land and the water can open an earned universe of wonders. But you must spend time in the villages, too, with the kindhearted people of the North to make sure you get it right. And you can't call the Arctic "the Far North"—it is "home" rather than "far" to the many people who live there.

So, after twoscore of Arctic journeys, in the summer of 2022, I'm on one more trip. I could not be on such an ambitious trip without all the previous experiences, which I'll use to sculpt a cohesive, whole, and fully developed sense of place. (Disclaimer 2: The more I learn, it sometimes feels like the less I know about the Arctic.)

But this time the agenda is different. I hope to better understand the climate crisis.

Chris and I are here to document it however we can. Since my first trip above the Arctic Circle in 1983, I have seen extraordinary changes in the landscape. Only three days underway and we've already flown over a wildfire to access our Walker Lake drop-off point. And yesterday we trudged underneath several bizarre, teardrop-shaped landslide thaw slumps—a.k.a. thermokarsts—caused by the permafrost thaw.

In much of Alaska, the Arctic Monitoring and Assessment Programme (AMAP) says that permafrost thaw from 2005 to 2010 has caused the ground to sink more than four inches, and in places to the north of us, twice that. The land collapses as the permafrost below it thaws like logs pulled out from beneath a woodpile. AMAP believes this will amount to a "large-scale degradation of near-surface permafrost by the end of the twenty-first century." This means that roads and buildings and pipelines—along with hillsides, Iñupiat homes, forests, and even lakes—will fall crazy aslant, or get sucked into the ground as if taken by an earthquake.

Our hoped-for documentation of the climate crisis isn't likely to be easy on this remote wilderness trip. We don't expect a picnic—known as Type 1 Fun to modern-day adventurers. A journey across the thaw on foot and by packraft for five hundred-plus miles won't resemble a back-country ski trip or a long weekend backpack on Lower 48 trails. We have planned for Type 2 Fun: an ambitious expedition that will make us suffer and give us the potential to extend ourselves just enough that there will be hours, or even days, that won't seem like fun until much later when we're back home. Then our short-circuited memories will allow us to plan the next trip as if nothing went wrong on this one. As if it was all just great Fun with a capital F. Ultimately, an important part of wilderness mastery is to avoid Type 3 Fun: a wreckage of accidents, injuries, near-starvation, or rescue. We've both been on Type 3 Fun trips that we'd rather forget.

I have more than a few outdoor expert friends who claim that the real trick of expeditions is always to be in control, regardless of calculated risk activity, and to avoid the proverbial "adventure." But I've never really succeeded. No matter how hard I try, on my trips in the North, things sometimes simply go south: someone forgets the stove, you fall and dislocate a shoulder, you forget to tighten the pee bottle in a crowded tent in a storm, you miscalculate a crevasse jump, you shower yourself with bear spray, or someone (not me) lets the tent blow away. You try to avoid these rookie moves, but sooner or later, particularly on a long expedition, the odds are that you or one of your partners will screw up, or the weather will thrash you, and you'll have an adventure.

Yesterday, in the first hour of our day, while we forded the Kalulutok, I tripped, fell in face-first, performed a splashy left-handed push-up in three feet of water, clambered upright—my packraft, clutched by its bowline like a leashed dog, yanked me backward in the current—and continued the hurried, bowlegged swagger necessary to stay upright in the river and keep my companion in sight.

No big deal (but it could've been a time-consuming raft chase if the bowline had slipped out of my hand). I just got soaked and it cooled me off.

As far as I could tell, Chris never tripped, much less paused, as we beat our way toward an unnamed pass several thousand feet above. Chris is a sponsored outdoor athlete, born thirty years after me, one of the world's most accomplished expedition kayakers, and a talented photographer and filmmaker. On first descents of the most hairball rivers on the planet, he films the guys who get made into heroes, then jumps in his boat and drops into the same maelstroms anonymously, a whitewater ninja with little of the credit or notoriety heaped upon his companions. He's kind, thoughtful, and dyed-in-the-wool laconic, so you have to pry to get him to speak a full paragraph.

We had planned the route with assurance from an acquaintance who had enjoyed an easy trek along the creek bottom to Walker Lake after a descent from the Arrigetch a couple of decades ago. Type 1 Fun. The caribou migration, a half-million strong back then, had punched game trails along the stream through dense alder thickets with a profusion of branches thick as human arms splayed in every direction.

We found something entirely different. The trails petered out a quarter mile from the lake, more than fourteen miles from the pass we had to reach. So we were forced to march directly up the slippery river with packrafts in tow. Or, as the river steepened and deepened, we teetered along banks strewn with rounded boulders beneath low-slung alder branches. While I struggled to catch Chris on day three as he made his way uphill through tussocks into a boulderfield above the choked creek, I figured out what had changed in this valley.

2 According to the National Oceanic and Atmospheric Administration's *Arctic Report Card 2018* "five herds in particular, in the Alaska-Canada region, have declined more than 90 percent and show no signs of recovery." There has been a 56 percent decline in migratory caribou and wild reindeer in the circum-arctic tundra regions—from 4.7 million to 2.1 million animals in the twenty-three herds monitored—over the last two decades.

The Western Arctic Caribou Herd population has dropped from a half million to 152,000 animals. This isn't news to wildlife biologists. Around the world, climate change and habitat loss have caused a decline in caribou populations.[2]

Then there's the Greening of the Arctic phenomenon, which will continue as we cross into the Noatak River valley in a couple of days. Basically, as temperatures warm and the summers lengthen, forests have begun to move north into treeless zones. Even the native shrubs—smallish alders, dwarf birch, and knee-high willows—have grown more than six feet tall and into dense thickets.

In stagger-step behind Chris (who has gained another hundred yards on me), I begin to get it. The Western Arctic Herd—shrunken since our acquaintance's Type 1 Fun trek through here a couple of decades ago—hasn't migrated through and left trails in the brush of the Kalulutok drainage, which has now turned into a lush bush-whacker's nightmare. As for the game trails we sussed out on Google Earth, the satellite pictures are probably several years old. Plenty of time for the Greening of the Arctic to transform the valley into a warmed-up arboretum.[3]

I finally catch Chris at a steep stream valley that leads up toward the pass. Reindeer lichen crunches and breaks in brittle white pieces beneath our feet as we sink several inches into the soft sphagnum moss. For the first time in three days, the heat has eased as the sun hides behind thick clouds propelled by a steady wind.

My hands, thighs, and calves have repeatedly locked up in painful dehydration cramps, undoubtedly caused by our toil with leaden packs in eighty-degree heat up the steep streambed or its slippery, egg-shaped boulders. After my water bottle slid out of an outside pack pocket and disappeared amid one of several waist-deep stream fords or in thick alders yesterday, I carefully slide the bear spray can (looped in a sling around my shoulders) to the side so it doesn't get knocked out of its

3 The article "Arctic Sea Ice Retreat Fuels Boreal Forest Advance," published in the February, 2024, *Science* 383, no. 6685, shows how open water caused by a lack of sea ice has allowed tree lines to expand across the North. This is also called "the Greening of the Arctic." Over four years, researchers studied nineteen sites in northern Alaska and found that the growth of trees and forest expansion is caused by and linked to open ocean water (through melting sea ice) associated with "warmer temperatures, deeper snowpacks, and improved nutrient availability." Then around the world, the article maps eighty-two different tree-line advances alongside more Arctic Ocean waters—including the Chukchi Sea, the Beaufort Sea, Hudson Bay, the East Siberian Sea, the Barents Sea, the Kara Sea, and the Laptev Sea—with ongoing sea ice loss. Like other studies of the Arctic climate crisis, greenhouse gases have caused a rippling domino of changes across various ecosystems.

pouch. Now, to slake my thirst, I submerge my head in the stream like a water dog. I'm beat.

To get Chris, a caffeine connoisseur, to stop, I simply utter, "Coffee?" His face lights up as he throws off his pack and pulls out the stove. I pull out the fuel bottle. Since Chris isn't a conversational bon vivant, I've learned not to ask too many questions, but a cup of coffee will always stimulate a considerate comment or two about the weather. As I fire up the trusty MSR stove with a lighter, we crowd around and toast our hands over the hot windscreen as if it's our humble campfire. We're cold and wet with sweat and we shiver in the wind. But at least we're out of the forest-fire smoke—this summer more than three million acres would burn in dried-out Alaska.

We slogged nine miles yesterday, but with constant route decisions and swift streams to cross and prolonged zigzags and thick brush to muscle through, the labor weighed on our backs like a thirty-mile day on trails. Still, I suspect that Chris's brief laments about the packhorse work are only empathetic commiserations for my exhaustion. Several months before the trip, as I trained with hill runs and bike rides, I emailed my younger partner and asked if he could, for my sake, gain fifteen pounds or take up cigarettes.

Yesterday on another well-earned coffee break, I zipped up after I'd watered the alders with a dehydrated, amber stream of urine. Chris sat in the heat, shirt off, with his back to me and I noticed his unusual anatomy: two latissimus dorsi muscles bulged like a set of transplanted quadriceps. I knew then that if this bushwhack-in-the-river-style backpack seemed an uneven catch-up race for me, once we took out the packrafts and he shifted his paddle muscles into gear on the Noatak River, I'd never catch him.

Today, with the all-day uphill climb and inevitable back-and-forth route decisions through the gorge ahead, we'll be lucky to trudge even five miles to the lake below the pass. *Why*, I ask myself, as Chris puts on

Emerged from the thick alders, Chris walks with camera equipment under one hand and a fully assembled paddle at the ready in the other, still days away from the Noatak River float. JON WATERMAN

his pack and shifts into high gear, *could we not have simply flown into the headwaters of the Noatak River instead of crossing the Brooks Range to get here? One step at a time*, I think, as I heave on my pack and wonder how I'll catch Chris, already far ahead.

Shards of caribou bones and antlers lie on the tundra as ghostly business cards of a bygone migration, greened with mold, and minutely chiseled and mined for calcium by tiny vole teeth. We kick steps across a snowfield, then work our way down a steep, multicolored boulderfield, whorled red and peppered with white quartz unlike any rocks I've seen before. As rain is shaken out of the sky like Parmesan cheese from a can, we weave in and out of leafy alder thickets while I examine yet another fresh pile of grizzly feces. I stop to pick apart these scat—and thumb through stems and leaves and root pieces—this griz appears to be on a vegetarian diet.

"Hey, bear!" We yell again and again until we're hoarse. I hold tight to the pepper spray looped over my shoulder to keep it from grabby alder branches.

On the first day of our bushwhack, an alder branch knocked my can of pepper spray out of its holster, and Chris and I searched for a half hour until we found the can half hidden by an alder trunk. In another tight tangle along the way, Chris lost his ski pole but didn't realize it until miles farther on. Minor mishaps all, but part of a score that might tally up into a so-called Adventure.

The stream gorge cliffs out at a waterfall amid boulders laced with spiderwebs so intricate I couldn't help but wonder if these Arctic arachnids were out for game bigger than mosquitoes. We climb out on the opposite bank and work our way along the cliff edge. But a half mile farther the route dead-ends, so we're forced to descend into the gorge again. With Chris twenty yards behind, I plunge step down through a near-vertical slope of alders and play Tarzan for my descent as I hang onto a flexible yet stout branch and shimmy-swing down a short cliff into another alder thicket. A branch whacks me in the chest and knocks off the pepper-spray safety plug. When I finish my swing to the ground, I get caught on another branch that depresses the trigger in an abrupt explosion that shoots straight out from my chest in a surreal orange cloud. Instinctively I hold my breath and close my eyes and continue to shimmy downward, but I know I'm covered in red-hot pepper spray.

When I run out of breath, I squint my eyes, keep my mouth closed, and breathe carefully through my nose, and scurry out of the orange capsaicin cloud. Down in the boulderfield that pulses with a stream I open my mouth, take a deep breath, and yell to Chris that I'm okay as I strip off my shirt and try to wring it out in the stream. I tie the contaminated shirt on the outside of my pack and put on a sweater. My hands prickle with pepper.

"You really okay?" Chris asks, concern on his face.

"Yeah. If that spray had hit me in the eyes, it would've been really, really bad news."

Then we're off again. As we clamber up steep scree to exit the gorge, my lips, nasal passages, forehead, and thighs burn from the pepper. As I sweat from uphill exertion, the pepper spray spreads from my thighs to my crotch like a troop of red ants, but I can hardly remove my pants amid the storm clouds and wind.

With the last of the alders below us, we enter the alpine world above the tree line. As I walk, slicked with peppery sweat, I understand—maybe even sympathize with—how bad a sprayed grizzly would feel, with its huge snout and a sense of smell so superior to my inadequate nose, which feels as if I've snorted Tiger Balm.

The final climb to the lake is up a steep streambed, with wafer-thin snowfields that creak under our steps. If the thin bridges break under our weight, it'll be a wet epic to clamber out of the icy stream with its slimy rocks. So, I pretend it's a perilous glacier and walk with light, delicate steps and distract myself from the potential crevasse plunge while I study hundreds of marmot turds strangely clung to the rocks like squashed button mushrooms gone black.

Although I have learned a lot on past trips, I am drawn back to the Arctic again and again. I am fascinated with a place that alternately resembles the lost world of the Pleistocene—gouged throughout by the ancient push of glaciers—then in the next moment is unexpectedly filled with weird spiders huddled on the side of their massive webs. When you really need distraction, and you pay attention to your surroundings, wild creature signs can appear like an overstuffed bazaar filled with feces of all sizes and shapes, story-filled track paths, and fur caught on a branch where an animal scratched its back—a moose, or maybe a caribou. Or so I occupy myself as I labor steeply uphill with an unwieldy pack and pepper-sprayed skin.

There's nowhere else to walk in this steep valley. The tundra trail alongside the stream is maintained by a grizzly as if it's a regular commute, given that the rest of the valley is flooded with sharp-angled boulders. Every hundred yards or so we spy more excreta as big as horse piles on matted tundra compressed several inches from the bear's weight. Its prints are longer than our size twelves and twice as wide. To make sure I don't lose Chris, I quickly punch open the scat piles with my ski pole and find them sun-crisped on the outside, with no bones inside.

These intelligent bruisers are commonly thought to be voracious meat eaters. But the truth is more complex: since animal prey is hard to find and harder to kill, bears have an uncanny nose for proteins found in any number of plants—peavine roots, *Boykinia*, and sedges—that I would do well to add to my nutrient-poor salads at home.

By the time we reach the lake, the drizzle has become a steady rain. I'm nauseous from the day's exertions and overheated underneath my rain jacket with the red pepper spray that I wish I had saved for an aggressive bear instead of a self-douche.

Camped alongside the roar of Kalulutok Creek in a bug-quelling wind, we can see our side-valley route to the unnamed pass, left of the mountain, that will take us to the Noatak River headwaters. CHRIS KORBULIC

A pair of Arctic loons ply the water but, loathe to be disturbed in their wild, inviolate basin, they beat their wings in double time and imprint dimpled tracks across the lake to get their fish-fattened bodies aloft and away from us. A bevy of snow buntings drifts out across the boulders like a white squall.

There's no sign whatsoever of human visitation, but I can imagine an ancient Iñupiat hunter could've padded through as stealthily as a wolf in his caribou mukluks as he contemplated this world apart from the river valleys on either side. He would've beheld the peaks all around and the strange, gargantuan boulders furry with the green and red glow of lichen and the lake that stares out from the middle of it all like the lone cerulean eye of the gods. Even without sun, it is a wonderland of delicately fringed, fuchsia-tinted *Diapensia* and bitter blue lupine stalks that provide painted contrast to the drab-gray, broken cliffs and granite slabs—macerated and jumbled-up when gargantuan tectonic plates arose from a shaken Earth to create the Brooks Range 150 million years ago.

Atop wet tundra that feels like a sponge underfoot, we pitch the Megamid tent with a paddle lashed to a ski pole and guy out the corners with four of the several million smaller boulders from the reduction of tectonic litter that surrounds us. I feel old here next to my unstoppable partner, but amid this primordial wreckage of Earth, I am interminably young even if I can't cop to it after the afternoon's rigors.

I fire up the stove, boil the water, and we inhale four portions of freeze-dried pasta inside the tent. We don't second-guess this evening's departure from wilderness bear decorum to cook outside and away from the tent because it's cold and we're tired. Chris immediately heads out with his camera. He's anxious to get out of the capsaicin-spiced tent; his eyes are watery from just being within several feet of me.

I've been reduced like this before—wounded and exhausted and temporarily knocked off my game. So, I tell myself that this too will pass, that

I'll get in gear and regain my mojo. That maybe I can eventually get my shy partner to loosen up and talk. That we will discover an extraordinary new world—the headwaters of the Noatak River—from up on the pass in the morning. And that I will find a way to accept and withstand my transformation into a spicy human burrito.

Snow feels likely tonight. It's mid-July, yet winter has slid in like a glacier over the Kalulutok Valley.

I am too brain-dead to write in my journal, too physically wiped out and overheated in the wrong places to even think of a simple jaunt through the flowers to see the view that awaits us and take photos. I remain in the tent desperate for relief and wonder what statute of age limitations should be placed on this kind of expedition as I pull down my orange-stained pants and red underwear and grab a cup filled with ice water. I try not to moan as I put in my extra-hot penis and let it go numb.[4]

Type 2 Fun for sure.

4   Bear spray contains 2 percent capsaicin (made from chili peppers); self-defense sprays used by police contain 1.2–1.4 percent capsaicin, which inflames mucous membranes and causes temporary loss of vision if sprayed in the face. Neutralization of the hottest natural pepper, habanero, takes two hundred thousand Scoville heat units (a number considered "Extra Hot": SHUs are the number of cups of sugar water needed to neutralize a spice). Self-defense spray measures at one million SHUs (considered "Extremely Hot"); bear spray, three million. In the January 6, 2021, Capitol insurrection in Washington, DC, rioters used bear spray on the police (one of whom died after he got sprayed).

PART ONE

Schooled

Prehistoric Times – Present Day

Above the valley of the Arrigetch ("fingers of the outstretched hand" in Iñupiaq)—
a rock climber's paradise—looking at the granitic Maiden Peaks in 1985. Dënéndeh,
Kuuvuan KaNianiq, Iñupiat, and Gwich'in Nàhn ancestral lands. JON WATERMAN

An Arctic Primer

2.4 Billion Years Ago–2022

Here's what I've learned.

Amid the immense landscape north of the Arctic Circle, it's easy to become disoriented. Much of the region has no trees, buildings, or roads. In this netherworld, short of visual cues that give perspective, a gaze across one of its broad valleys creates confusion. A distant grizzly can look like a nearby ground squirrel. I've repeatedly mistaken sandhill cranes for caribou.

Cone-shaped hills that resemble mini-volcanoes in the distance are pingos, squeezed up from below like teenage Earth acne by an expansion of underground ice. Try to walk across most flat valley floors—that beckon you with the initial appearance of plush tundra carpets—and you can sprain an ankle in bogs filled with foot-and-a half-high, eight-inch-wide, mushroom-shaped, grassy tussocks. Too flimsy for more than a snowy owl to alight on.

Add in the frequent mirages that distort perspective along the coast and inland when warm land air mixes with the cool sea ice or snowfield air. Even in the heat of midsummer, rivers are often plastered with thick pans of overflow ice. Called *aufeis*, these winter-leftover plates of frozen river stack ten feet tall in outlandish, onion-like layers.

Sixty-degree days (I have repeatedly checked thermometers) are often weirdly raked by sea ice winds that make testicles retract, nipples shrivel, fingers grow numb, and demand a down jacket, hat, and gloves. Against a headwind on Arctic rivers, you'll fight blue-lipped and numb-toed all day to paddle a few miles.

And no Arctic veteran underestimates the mosquitoes. A hungry cloud can drive otherwise highly evolved caribou crazy as they jump, buck, and sprint back and forth in torment. Although there is a legend of an Iñupiat baby killed by mosquitoes, an adult passed out on the tundra in midsummer under a healthy cloud of these insects theoretically could be exsanguinated in a few hours. And it doesn't help to know that there are autogenic species of mosquito in the Arctic that can still reproduce and lay up to one hundred eggs without a blood meal.

Winter is omnipresent. You can sense the implacable mass of the gigantic Ice Age glaciers that filled the Brooks Range, shaped valleys, and crushed everything in their path. Then there are the abbreviated summers. The obvious freneticism of the animals in a flurry of constant movement as they work around the clock for survival and nourishment before the long winged or hooved migration south begins. Most of the animals, like us non-villagers, are transitory visitors.

For a couple of months in summer the sun doesn't set, and all night long, the birds compete in lusty chorales. Chris wore eyeshades in order to sleep.

The Arctic can be verdant in places and even look lush with ankle-high plants that flourish across summer-dampened permafrost. Yet the sparse annual precipitation makes many places arid. In the heart of the Brooks Range above 2,500 feet, Gates of the Arctic National Park and Preserve averages five to ten inches of precipitation a year, same as the Mojave Desert—with its sand dunes and sparse Joshua tree forests—in Southern California.

No more than a few feet below the surface, the earth is frozen and locked in ancient ice for hundreds of feet. A stark reminder—when you pass a recently eroded riverbank that shows its frosty innards like a jacked-open vanilla ice cream carton—of the Pleistocene's endurance. But now it's under the process of a great thaw.

Taken altogether, the Arctic is not an ideal vacation destination. There are grizzlies and polar bears that have killed and then hungrily consumed people. There are lion's mane jellyfish that ply the Bering Strait and the Beaufort Sea with neurotoxic tentacles longer than a blue whale. There are the bumblebee-sized, orange-striped nose bots, or warble flies, that routinely place maggots or eggs into the noses or onto the hair of caribou. They'll also settle for human hosts.[1]

But amid the utter strangeness, the Arctic is scented with aged spices: the sweet turpentine of Labrador tea and the lilac perfume of the tiniest, most delicate pink twinflowers amid real bumblebees (not botflies) that gather pollen. The place is ribboned with rivers finned by peculiar and beautiful fish, playful ground squirrels chirp throughout, and the midnight sun burnishes the hillsides gold and makes the frequent rain showers glow like lemonade poured out of the sky. Beneath it all the active layer of permafrost tundra presents as a giant live body and holds forth an abundance of blueberries, bunchberries, crowberries, lingonberries, salmonberries, cranberries, and bearberries. So I have learned to swallow a bit of discomfort. To take deep breaths and let the landscape open my heart and fire my imagination like no place on Earth.

As the sun sinks below the horizon in August and night returns, in a curious continuity with the brilliant summer light show, solar winds blow particles into the upper atmosphere. This enlivens the night sky with colorful evanescence: gauzy saffron clouds and beams of viridescent and roseate flickers. The mysterious northern lights were named (by the astronomer Galileo four hundred years ago) after the Greek goddess Aurora, who flew through the sky to announce the dawn, and Borealis, the god of the north wind. Beneath the lights on a cold, still night in the wilderness even jaded skeptics can become Arctic Believers.

1 While nose-bot infestations of humans haven't been documented, in the Arctic, warble flies have repeatedly caused larval infestation of human eyes—known to medical professionals as ophthalmomyiasis. Normally, warble eggs are laid on caribou fur; when the eggs hatch into maggots, they burrow under the hide and feed on the caribou's flesh under its back (as many as a thousand maggots could kill their caribou host). Nearly a year later, they tunnel back out of their host (with innumerable exit craters that pimple the caribou hide), fall to the ground, and turn into adult flies. They can fly up to fifty-four miles for a new host. In a medical journal article "Human Ophthalmomyiasis Interna Caused by *Hypoderma tarandi* [warble fly]," the authors document thirty-two cases of human infestation. Although the "parasite does not appear to complete its life cycle in humans," it was shown to cause pain and blindness in more than a few of its victims and grow up to three millimeters before being surgically removed or killed with antibiotics.

Grass of Parnassus, a.k.a. bog star (top), and the edible bistort (bottom). CHRIS KORBULIC

A fox skeleton amid ripening crowberries (right) in August, Chukchi Sea coast.
Iñupiat ancestral lands. JON WATERMAN

On a dark midnight in late August 1998, in a tent on the remote shore of the Amundsen Gulf in the Canadian Arctic, a hundred miles from another human, I had been awoken by what sounded like the rustle of a bear outside the tent. After a "Go away, bear!" yell, I wormed out of the tent and went prostrate on the flat, treeless tundra to try to identify the intruder in profile against the starlight. Although no creature moved within visible distance of the tent, the northern lights blazed above me in a brilliant green shimmery set of curtains that crinkled open and shut as if a bear panted and swished and pushed through the knee-high dwarf birch and sedges. No way could I go back to sleep.

Scientists have recently proved that you can hear the lights.[2] They've been described as two pieces of silk that are rubbed together. Or, as some people (me, for instance) claim, like an animal that stalks you through the brush.

While the assertion that the spectacle could be heard by human ears had long been derided as an old wives' tale, the observant and sensory-astute Iñupiat have always listened to the lights and ascribed the spectacle to spirits who kicked skulls through the sky. Iñupiat used to carry knives when the lights danced above, or threw dog feces and urine aloft for protection. Still, they tried to revere the ghostly spirits that played above.

It is a temptation to perceive the Arctic as a dreamscape, its logical conclusions edited into a fairy tale not bound by plot structure. After a trip to the North, it can feel as if you've awoken from a long night's sleep but can't remember the dreams that will offer connective threads to your life. I have repeatedly returned to the Lower 48 and found it hard to drive a car at more than thirty miles per hour. The outside world, as the Iñupiat say about the climate crisis and cities to the south, "is now moving faster."

The Arctic feels like a place that might offer redemption, whether you need it or not, even as it repulses you. When I go to the Arctic, I sense an

2 In 2016, a Finnish professor and his colleagues presented a widely accepted study (conducted in 2012) based on their audio captures of hundreds of possible "auroral sounds." Most surprisingly, the study suggested that the aurora borealis can be heard even without visible northern lights. The researchers also identified that the sounds came from 230 to 330 feet above the ground, as a shallow layer of cold air along the ground built electrical discharges beneath a higher layer of warmer air—all caused by the geomagnetic disturbances potentially triggered by the aurora borealis.

opportunity to recharge and lose the taints of lost love and miscommunications and life blunders. Up here, you can start all over again.

To make sense of it all, to feel less lost when I first went to the land above the trees thirty-nine years ago, I learned all I could about the animals. The names and habits of the many birds: the Arctic terns that wheel through the air, the rattle and bugle of sandhill cranes, the jerky stream dance of dippers. The culture of the Northern People. And how this top hat of the Earth had been built over the eons.

Late in my seventh decade, I came to realize that I dove deep into the Arctic for an orientation absent from my life. I went to find gravitas and value in a place that concurrently frightened, fascinated, and touched me with its vastness, abundant light, and mountains-cum-relics of a long-vanished sea.

I also sought mastery and wanted my journeys through the Arctic to release me from my moribund introversions, as if the grandeur of what the Iñupiat refer to as the Great Earth and its Weather would somehow confer beauty and peace to my soul.

I learned that if I wanted to get outside my head, it was possible to develop a sense of place and come to see the Arctic as an entity of its own, so far removed from the Lower 48 as to be another country—if not another world, a timeless one of evolutionary phenomena, where the ancient past sits with the present landscape.

Now, as the wonders of the Arctic undergo aberrant change, this knowledge of place has become more important. Even essential. As the global heat-up alters the migrations of the fish, mammals, and birds, the Iñupiat are also affected. The sea ice has melted away as storms erode shorelines and flood villages. Forests are slowly on the move north along with animals new to the Arctic. The permafrost has begun to thaw, and lakes have disappeared as riverbanks and mountainsides droop like frozen spinach left out on the counter. Unprecedented lightning storms have brought wildfires that sweep across tundra dried out

in weird heat cycles that haven't existed in the Arctic for hundreds of thousands of years. To call it simply "change" or think that nature or its inhabitants can endure is a naïve estimate of the inevitable crisis.

So, it's high time that we truly understand the Arctic. Lest we forget what it once was. Lest we forget …

———————

You can imagine most-ancient Alaska when you slog past a dark boulder embossed with whitened squiggly patterns of shells and dendritic arms that show how the region once lay under a shallow sea, broken by low rounded mountains. Back then, three hundred to four hundred million years ago, before the seven continents existed, life consisted of spineless invertebrates: coral, tiny-tentacled and shelled brachiopods, and the now-extinct, three-lobed trilobites.

In a museum with the boulder as a canvas, the fossils could be perceived as minute art forms of remarkable symmetry, but on foot in the vastness of the far-flung Brooks Range you want to feel the sense of purpose in it all and understand eons of change. The fossils represent some of the earliest life on Earth, a glimmer of a miraculous evolution of single-celled organisms that fused into many-celled primitive animals as their body plans diversified and radiated. Eventually they exploded into an array of invertebrates that would someday mutate into larger creatures—millions of caribou, for example—constructed with spider-webbed catchments of neurons wired to lobed brains that would transform them into sentient souls. Or so you can imagine.

To put it all in perspective, to wrap your arms around climate changes even further back in time, Earth had already experienced several ice ages. The Huronian glaciation occurred 2.4 to 2.1 billion years ago and was likely caused by a lull in volcanic activity. Then 720 million years ago, amid the aptly named Cryogenian period, much of the planet became encased in snow and ice[3] as the greenhouse effect weakened and carbon dioxide was pulled out of the atmosphere. More ice ages

3 Based on evidence of ancient glaciation at the equator, NASA scientists believe there were two "Snowball Earth" glaciations in the Cryogenian period that each lasted about ten million years. Amid these massive glaciations, temperatures dropped as much as 12 degrees Celsius below zero in a time when the sun's strength had declined about 6 percent. Since there is no evidence of wholesale extinctions that should have been caused by frozen oceans, an alternate "Slushball Earth" theory contends that the oceans weren't completely frozen around the globe, and this allowed marine life on the planet to continue.

The sight of a huge barren ground grizzly's tracks—let alone the bear itself—is enough to give pause to even our large group in the Noatak River headwaters, August 2021. Dënéndeh, Kuuvuan KaNianiq, Gwich'in Nàhn, and Iñupiat ancestral lands. JON WATERMAN

followed, amid millions of years of incredible plant growth and periods so warm that no ice existed even at the poles.

The world heated up 252 million years ago, which caused the largest die-off of plant and animal life. In this era before dinosaurs, the Permian extinction—an extended era of climate change that might mirror the human-caused crisis that started with the Industrial Revolution—killed an estimated 96 percent of marine life as oceans lost oxygen and the invertebrate creatures (such as trilobites, brachiopods, and corals) and algae piled up on the seafloor as a dense foundation of corpses.[4]

Reptiles that survived the heat and dryness of the Permian extinction evolved into dinosaurs. Just over two hundred million years ago much of Earth's climate remained tropical, and oceans—like the shallow sea that overlaid the area later to become Alaska—held an abundance of life. There were sea turtles, plesiosaurs that resembled giant newts with four flippers and an elongated neck, and ichthyosaurs. (In 1950, geologists found an ichthyosaur fossilized in the western Brooks Range[5] that resembled a giant reptilian dolphin.) Legions of small fish swam amid an abundance of microscopic plankton and other minute life-forms. As creatures microscopic and gargantuan died over thousands of generations, they accumulated and were compressed into an incalculable tonnage on the seafloor, atop the dead foundation of invertebrate life from the Permian extinction. Ancient life would, in turn, be compacted into more limestone or squeezed into coal and petroleum beds.

One hundred and fifty million years ago a massive upheaval began. The northern edge of the 140-million-square-mile Pacific tectonic plate—80 to 120 miles below the seafloor's necropolis of pulverized life-forms—began to collide with the lighter North American Plate.

It happened. Inconceivably. Slowly.

Pushed a few inches per year, the saturated rocks of the Pacific Plate bulldozed under the North American Plate. If photographed with underwater, time-lapse photography, the seafloor would have appeared

4 University of Washington and Stanford University researchers have theorized that the lack of oxygen in the world's oceans amid this "Great Dying" had been preceded by global warming and increased acidification—startlingly similar to twenty-first-century changes. Yet, unlike today's human-caused climate crisis that took little more than a century to heat up the Earth, the global warming of the Permian extinction had been caused by Siberian volcanic eruptions over millions of years. "This study," one of the researchers said, "highlights the potential for a mass extinction arising from a similar mechanism under anthropogenic [human-caused] climate change."

5 In 2002, paleontologists encased the twenty-five-foot-long, 210-million-year-old ichthyosaur in plaster and the Army flew out the one-ton package with a Chinook helicopter to Fairbanks. In its digestive tract they found fish bones, eaten when the Brooks Range lay beneath the sea.

as if it were pushed by a humongous underground being, headed for air in a slow-motion ascent.

It took tens of millions of years.

Finally, Alaska arose, breached the surface, and shrugged off the shallow Arctic Sea. As the dense Pacific Plate continued to submarine at a forty-five-degree angle under the lighter North American Plate, it tore up 2,200 miles of ocean floor.[6]

Like Alaska, life started in the seas and included the ancient fish that crawled out of the surf and created the evolutionary lineage of terrestrial creatures great and small. Although the perplex magnificence of Alaska could also be perceived as a divine creation, the tectonic forces that built it all—bit by bit, as mountains lifted from north to south—were a decidedly more haphazard and violent affair. As the most northerly extension of the Rocky Mountains, the Brooks Range arose as gigantic, mangled pieces of Pacific and North American Plates still engaged in a planetary rugby match.

The friction created a subterranean blast furnace of heat that melted rock and earth and pushed it a mile and a half skyward along with the scrum. Mixed into the volcanic rock found throughout the Brooks Range is that several-hundred-million-year-old limestone—the saltwater foundation of ancient corpses piled on the seafloor from the Permian extinction, buried, and further compressed by the great earthen brawl. Hard and compact as cement, the limestone resisted erosion and stood until modern times as a testament to the passage of the ages.

To slowness.

Atop the Arrigetch Peaks, limestone spires pointed their sharpened middle fingers into the sky. As if the ancient life, unfazed by later glacial erosion, told the ice to go plow some looser mountain.

Terrestrial life flourished. North of the Brooks Range, scores of different dinosaur species roamed the lush river valleys. Several-ton duck-billed

6 Today this is known as the fifty- to one-hundred-mile-wide Aleutian Trench, which drops more than five miles deep and shows the unfathomable forces generated by a head-on collision of the planet's crustal plates.

High up in the milky blue Noatak River headwaters, filled with fine particles of glacial silt, Chris passes by a riverbank with white permafrost ice exposed like a jacked-open vanilla ice cream carton. JON WATERMAN

hadrosaurs walked on two legs amid placid, horned ceratopsians reminiscent of modern-day rhinos. The six-foot-tall, skinny Troodon used plate-sized eyes to hunt in the dark and avoid theropod carnivores like the twelve-foot-tall tyrannosaurs, which couldn't catch and eat the faster, even fiercer dromaeosaurids, cousins of the velociraptors.

For less than one hundred million years throughout the greenhouse conditions of the Cretaceous period, the Arctic had been too warm to support ice caps. Dinosaurs ruled. Then sixty-six million years ago a nine-mile-wide meteor collided with the Yucatán Peninsula in Mexico and blasted out a crater ten times as wide. Even the great beasts of the far-off Alaskan Arctic that survived the initial blast and the tsunamis that swept the rest of the world would quickly perish when they breathed in the sulfuric aerosol haze. The dust that wouldn't settle for years caused the sun to dim and temperatures to change. This event killed 75 percent of all species on the planet. Like past cycles in Earth's tumultuous history, year-round winters followed year-round summers—as if a steroidal icebox repeatedly got unplugged.

About thirty-four million years ago, global temperatures began to drop precipitously due to a drop in atmospheric carbon dioxide. The polar ice caps reformed. With only brief spikes of warmth that would repeatedly thaw waters to the North Pole, the cooled-down Earth would prevail. This most recent refrigeration climaxed three million years ago as we (or rather, our Australopithecus ancestors) walked upright in a glacier-free Africa. Elsewhere, ice sheets eventually covered a quarter of the planet's land area and sea levels dropped several hundred feet.

Glaciers in the Brooks Range dammed the principal Noatak River drainage. The glaciers held back a series of lakes, collectively called Noatak Lake by geologists. It stretched 1,700 square miles. The sheer weight of the glaciers physically depressed the land and lowered Howard Pass (now 2,200 feet above sea level) a thousand feet. For several millennia, the westward flow of the Noatak River shifted north out of Noatak Lake and over the Brooks Range through Howard Pass.

Water vapor and rain—when it warmed up enough for liquid precipitation—froze most everywhere north of what would become the Canadian border. Temperatures in the Arctic plunged as low as minus 80 degrees Fahrenheit. Snow fell incessantly. Since more snow fell in winter than melted in summer, glaciers grew and began to creep, blob-like, under their own weight and pressure (the dry North Slope of the Brooks Range remained free of glaciation).

Over the millennia, the ice at the bottom of these mega Brooks Range glaciers would be compressed into what resembled rock, composed of the crystalline form of water. The bulletproof ice plowed beneath the limestone peaks and polished the lower layers of granite alongside the iron-red rock and outcrops of super-hard quartzite now seen throughout the range. Like a giant anvil dropped onto a wild lawn, the glacier crushed and suffocated the live mat of earth below.

As the planet's water was locked up into Canada-sized ice sheets elsewhere on the continent, many rivers came to a halt. The Beringia landmass—at first a series of islands—would begin to open between Asia and Alaska. At the same time, the ground beneath all this ice on either side of the land bridge (and even on the seafloor) would freeze solid for hundreds of feet below the surface,[7] the origin of today's permafrost.

One can't understand the Arctic, or the climate-change thaw, without a careful study of permafrost, the shape-shifter of all land in the Arctic. Iñupiat call the ground *nuna*; when the upper, active layer of ground refroze every fall, they would say, "*Nuna qiqitkaa.*" To refrigerate their meat, they simply dug deep into the nuna to build an ice cellar (*siġluaq*). Since the ground was always frozen several feet down below the active layer, to name their cold storage otherwise would have been a fool's repetition. They wouldn't call ice "cold." But in recent years villagers can't rely on siġluaqs because the permafrost thaw floods the cellars.

Since frozen Siberia has long been inhabited, Russians started to study the "eternal frost" (*vechnaya merzlota*) more than two hundred years

7   It was likely that permafrost had been created, then thawed, amid ancient ice age cycles hundreds of millions of years ago. Many scientists believe that today's permafrost could have been created three million years ago when the most recent Ice Age started, but so far, the oldest permafrost dated is in Siberia, at 650,000 years old, and in the Yukon, dated at 740,000 years old.

8   What we've learned since then, with knowledge from the Russians, is that permafrost underlies 15 percent of the ice or glacial-free zones of the Northern Hemisphere (or 11 percent of the globe's surface, which includes Antarctic lands and high alpine mountains outside the Arctic). Eighty percent of Alaska is underlain by permafrost. This includes the most northerly continuous zone (29 percent of the state—essentially above the Arctic Circle), the discontinuous (35 percent), and sporadic or isolated (16 percent) zones of permafrost. Although permafrost is mostly absent beneath large lakes that don't freeze solid, there are regions of the Arctic Ocean—with sea bottoms exposed in the Ice Age—that have permafrost, now known to be in active thaw as ocean temperatures warm.

ago. A century ago a scientist called vechnaya merzlota the Russian Sphinx and analogized it as a great mystery that needed to be solved so that wells could be dug, roads built, and houses constructed.[8]

To be considered permafrost, the ground must be continuously frozen, below 32 degrees Fahrenheit, for at least two years. The exception is in summer, the top active layer of the greater permafrost mass thaws several feet down, and below, the ground is either dry, frozen dirt, or layered with ice. Over the eons, during warmer times, the ground covered and preserved trees, dinosaurs, mammoths, and volumes of plant life. Then it all froze solid. If, and when, the ground thaws again, microbes within the dead plants and animal carcasses will start to break down the organic matter and release several million years' worth of locked-up carbon and methane into the atmosphere.[9]

Scientists liken the permafrost thaw to the parable of the blind man who tries to describe the size of an elephant. No one can accurately predict how much of this gargantuan underground repository of carbon will be released into the atmosphere. Or how much it will warm the planet. But to villagers who have seen the thawed permafrost warp or collapse their houses, the crisis has already begun.

Permafrost makes the Arctic that surrounds the villages into a mostly treeless, open, extremely corrugated plain. In places, however, tree roots penetrate the top active layer but are unable to stand tall in the frigid air, so the stunted spruce or dwarf birch grow low to the ground (that is, until the Arctic begins to warm). In summer as the permafrost active layer thaws, trees have to adapt to the saturated ground that squishes underfoot like a bog.

The shapeshift nature of the ground is caused by the freeze and thaw of the permafrost active layer that, over long periods, can, among its other tricks, push rocks up to the surface. Movement of water in the active layer, along with the freeze-thaw cycle that occurs in summer, creates small tussocks, mounds, extended sections of remarkably similar

9 As per research that pooled observations from more than a hundred Arctic field sites, published by a team of over fifty international scientists in the journal *Nature Climate Change*, permafrost released about 1,662 teragrams (with each unit equal to a trillion grams) of carbon each winter from 2003 to 2017—double that of past estimates. In summer, plants absorb less than two-thirds that amount of carbon, and that allows more than a third to escape into the atmosphere and increase temperatures through greenhouse gas effects. While these emissions have not increased since 2003, the researchers believe that a permafrost loop, which creates a greater thaw, is already underway. The study did not mention the permafrost's release of the more potent greenhouse gas, methane, which contains one atom of carbon and four atoms of hydrogen.

patterned polygonal ground, sinkholes, caverns, tunnels, huge ravine-shaped slumps, and pingos (more than 1,500 in Alaska) up to six hundred feet high. It's as if giant wet blankets had been shaken and then laid down and frozen into place with terraces, folds, and high points.

Hotshot workers who fight the now-frequent wildfires in northern Alaska often reverse the chains on their chain saws and cut backward through the soil's active layer, creating permafrost ditches to stop the fires. Although these icy fire lines are more effective than lines dug in normal soil, exposure to warm air causes more slump features and permafrost thaw and leaves behind water-filled trenches for decades afterward.

If, in summer, a section of Arctic tundra were to be cleanly quarried down to six hundred feet and then measured immediately (before exposure to the air warms it up), the temperature of the first few soggy feet of the active layer would register in the high thirties to low fifties. Then, as you go down to either the frozen dirt or sections of white ice below the active layer, the temperature would eventually reach 32 degrees Fahrenheit—until a point known as the level of zero amplitude, which occurs anywhere from ten to forty feet below the surface. Below zero amplitude, the temperature warms to 50 degrees until the permafrost ends at unfrozen ground warmed by the Earth's internal heat. If measured in winter, the active-layer temperature would be less than 32 degrees (dependent upon the air temperatures) down to zero amplitude, where it would hit 32 degrees again.

Tens of thousands of years after permafrost had been created across the North, glaciers grew across the continent. The shallow waters of the Bering Strait between Asia and Alaska began to recede. The dry land expanded, while Earth's water continued to freeze, and continental ice sheets grew as high as continuous, cemented-together mountains.

Beringia was first exposed 35,700 years ago. It didn't flood back over for twenty-five thousand years—just about the time that glaciers began to recede in the Brooks Range. Fossil evidence shows that Beringia

The finely sculpted, streamlined form and iridescent blue scales of a grayling, with its distinctive, sail-shaped dorsal fin. You wouldn't want to dismiss the presence of their soul. Grayling are threatened by streams warming throughout the North. Iñupiat ancestral lands. JON WATERMAN

allowed plants and animals and humans to migrate between Asia and America.[10]

While the biblical epic of Noah and the great flood can be seen as a cogent (albeit fictional) allegory about a past era of climate change and extreme storms, its real-life equivalent is Beringia. Animals and humans used the exposed landmass to cross from one continent to another. It lasted for thousands of years amid a period of intense glaciation in North America that left northeast Asia high and dry.

One of the many Beringia migrants was the camel, which first evolved in North America forty million years ago. A three-and-a-half-million-year-old fossil from Arctic Canada shows that the thick-furred camel stood twice as tall as modern camels and utilized humps for fat storage in the cold, along with wide, flat feet to walk in the snow. These predecessors of the modern camel migrated across Beringia to Asia and, eventually, to the Middle East—just before it went extinct in North America 10,000 years ago, along with nearly seventy other species.

Immigrant megafauna like the elk came from the opposite direction and populated the continent from herds in Siberia. Human hunters followed the prolific game on foot, and later in boats, as Beringia flooded back over 11,000 years ago. But for all the traffic that theoretically occurred over thousands of years, there is a huge gap in the prehistoric record. The first sure record of human habitation in the Alaskan Arctic didn't occur until 12,000 years ago.

Two billion plus years of geologic and human history show that it often took at least tens of thousands of years for the Earth's climate to repeatedly change. Which caused massive die-offs—such as the Permian extinction that cooked the planet. If humans had existed then, they would have perished with the trilobites.

Back then, it all happened. So. Slowly.

10 Beringia is also referred to as the Bering Land Bridge. More than just a bridge, it stretched out as a 1,000-square-mile landmass that now underlies half of the Bering Sea. At the Glacial Maximum, the land called Beringia measured more than 600 miles from the Arctic Ocean south to the Aleutian Islands. It's theorized that early hunters built villages on this ancient Atlantis, amid a high-plains grassland rife with animal life.

Until humans began to arrive in places like Alaska. And soon enough, our species quickly jumpstarted a new period of extinction. Since 1500, 881 animal species have gone extinct (due to pollution, development, and climate change—according to the International Union for Conservation of Nature). This explains why the Iñupiat say that "the Earth is moving much faster now." Compared to repeated prehistoric climate change (except for the meteorite that hit sixty-six million years ago), the pace of climate change since humans began to migrate across continents—mostly after the recent Industrial Revolution—has already begun with ominous meteoric speed.

Before human-caused climate change, one can best picture the ancient Arctic migrations from the Iñupiat village of Alaska's Little Diomede Island (population 100). From there, it's a little over two miles west over the wind-frothed, icy waters of the Bering Strait to Russia's hilly Diomede Island (which fast-forwards the clock twenty hours across the International Date Line). And one of the best ways to understand the immensity of ancient Brooks Range glaciation is to float the Noatak River. Past its permafrost-exposed banks, you can try to gauge the distance across the glacier-plucked valley. Once afloat, amid the thaw and extinctions of the past, it's not hard to imagine your kayak as a time machine. My journey into Arctic wonder began there thirty-nine years ago.

The Noatak River
August 1983

Trip Plan: *On my first trip to the Arctic, we flew from the village of Bettles in a floatplane to Pingo Lake alongside the Noatak River. We planned to paddle forty-five miles downstream to Lake Matcharak, where another plane would pick us up a week later. I was officially on patrol as a park service ranger, on loan from Denali National Park. But those seven days on that prepossessing river—in the heart of Gates of the Arctic National Park—stretch as a long and idealized vacation in my memory.*

I had never been anywhere like it. We were surrounded by snowy peaks that could have been a mile or perhaps five miles off and—when we pulled our paddles out of the river to listen—the aqua-blue Noatak River tittered against the hull of our tandem kayak with unseen and minuscule grains of silt washed down from glaciers hidden above.

To call this place pre-Columbian—as if it simply preceded the colonies—would be a slight to culture and landscape. They span, respectively, back to the Pleistocene and the Mesozoic.

PREVIOUS SPREAD: During the autumn migration, over a hundred thousand of the Western Arctic Caribou Herd swim rivers like the Alatna. Climate change and habitat loss have caused their numbers to plummet—like most caribou herds throughout the Arctic. SETH KANTNER

I couldn't help but feel unsettled, even reduced, by the immense sky and landscape that surrounded us. The map didn't always help since the glacial river valley stretched so wide with such ice-rounded and flattened-out features that it proved hard to pinpoint our location. A hundred miles' worth of Noatak River shown on our small-scale map had been reduced to a shrunken curlicue of millimeters. Nor did any of the many wide river meanders seem to match up with where we thought we were when we could locate ourselves on the map.

This was our second day on the river. Ranger Dave Buchanan and I had packed vital tools: a shotgun for bear protection, journals (while he used his notes to write a park service report, I resurrect mine here), and a pair of park service–issue binoculars that Dave wore around his neck. The binoculars were vital to identify strange, minute objects in the distance without the perspective and visual cues of trees (the upper Noatak is all tundra), or people (there were none), or human trails (zilch), or buildings, signs, and familiar landscape features (all absent). The binoculars that Dave passed to me showed that what appeared to be a distant grizzly turned out to be a small brown cliff face fifty times bigger than any bear because a pinpricked-sized bird in the sky—maybe an eagle—gave the whole scene needed perspective.

Bear, Jon? Dave asked.

No, definitely not.

The scale of it all seemed too much to process. If you were unprepared for this vast and isolated valley, with such a momentous sky, it would be easy to feel overwhelmed. Still, if you studied things carefully, if you got down on the ground and examined the dozen different plants that would fit in your hand, you would find the Arctic to be a cornucopia.

The great naturalist John Muir visited the Noatak delta in 1881 and wrote: "The most extensive, least spoiled, and most unspoilable of the gardens of the continent are the vast tundras of Alaska. . . . As early as June one may find the showy *Geum glaciale* in flower, and the dwarf

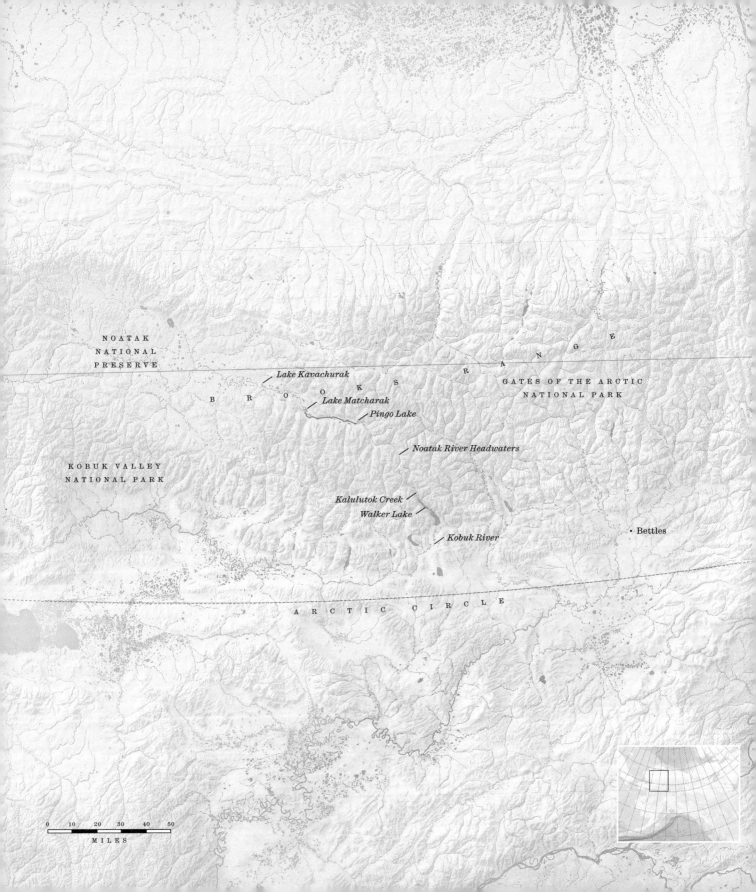

NOATAK
NATIONAL
PRESERVE

GATES OF THE ARCTIC
NATIONAL PARK

B R O O K S   R   A   N   G   E

Lake Kavachurak

Lake Matcharak

Pingo Lake

Noatak River Headwaters

KOBUK VALLEY
NATIONAL PARK

Kalulutok Creek

Walker Lake

Kobuk River

• Bettles

A R C T I C   C I R C L E

0 10 20 30 40 50
MILES

willows putting forth myriads of fuzzy catkins." He then listed three dozen more plants, with their "bright stars and bells in glorious profusion, particularly Cassiope ... the most abundant and beautiful of them all." He waxed on for several more paragraphs about other plants and wonders of the place.

But when the wind blows through one's layers amid stealthy grizzly bears and fierce clouds of mosquitoes, it's easy to feel undone. This proved true for S. B. McLenegan who trudged and paddled up the Noatak River in late July 1885. McLenegan was no John Muir. "The landscape was one of the bleakest imaginable," he wrote. "Not a sign of life was anywhere visible, and the cold piercing blasts which swept across the tundra caused us to realize keenly the solitude of our position and only increased our desire to see the end of the journey." He and a single companion fled back to the delta in an Iñupiat sealskin boat by early August as the snow began to stick.

In the distance behind our aluminum-framed, rubberized-nylon kayak— a boat composed of more than three dozen parts when we assembled it at the put-in—arose a hollow, chesty bugle call: *Kar-r-r-o-o-o, Kar-r-r-r-o-o-o, Kar-r-r-r-o-o-o*. Then the noisemakers came into view, and we craned our heads back to watch as two massive birds rounded the river bend to fly directly over our heads. Their stretched-out necks and the sound of their wingbeats—like leathery fans that scrunched air— brought to mind pterodactyls.[1]

"Sandhill cranes," said Dave, my companion in the stern. "Heading south."

We listened until the prehistoric croak calls died out downriver, and then there was only the silt that continued to abrade the stretched-tight rubber hull of the leaky, government-owned Folbot like swipes of light-grain sandpaper.

Dave worked as a backcountry ranger for Gates of the Arctic National Park and Preserve. My trip required a flight from Fairbanks to the village of Bettles, where Dave lived, then our hour-long floatplane flight

[1] The sandhill crane stands a meter tall and its direct descendants have been on Earth for more than two million years. One ten-million-year-old fossil of a big bird—that likely flew over Beringia—has the same structure as the modern sandhill crane.

Our forty-five-mile route in 1983 in the Noatak Headwaters of Gates of the Arctic National Park. Along with Noatak National Preserve, the two park units share thirteen million acres of legislated wilderness in the largest, undisturbed river basin in America.

2 Since we were still in a relatively
technologic dark age, satellite phones
did not yet exist. It took another
fifteen years for the first unreliable
sat phones to be released. I used one
once, and when its battery died,
never again. I preferred to stick to
the principles of true self-sufficiency
and solitude, measured against
the small chance that I might need
a rescue. Also, I always traveled
as light as possible without any
technological gadgetry.

into the Noatak. To walk into the park from the Dalton Highway north
of Fairbanks, many miles across the soggy and buggy muskeg, would
not be "a walk in the park." Only a few modern-day frontiers-people pull
it off without an airplane, mainly because the walk-in demands a huge
time commitment that most people would rather spend inside the park.

I first met Dave nine years earlier in the New Hampshire mountains
when we rescued a man who had swallowed a bottle of sleeping pills
and carried him down a steep trail to an ambulance in time to be
resuscitated. Two years later Dave and I went to Alaska, where we
attempted to climb Denali. Back in New Hampshire, when he left his
Appalachian Mountain Club caretaker job at Tuckerman Ravine on
Mount Washington, I took his place. A half dozen years later, in 1982,
I would attempt to fill his double boots as a ranger on Denali, while he
moved up to the Arctic.

Dave was serious and dedicated to his work on the river. I initially perceived park service patrols as a boondoggle until the Arctic opened my
eyes. Dave helped set the hook for what would become my lifetime fas-
cination with the Arctic. After my fourth season with the park service,
I abandoned a government career and glacial mountains for remote
wilderness trips that allowed me to immerse myself in northern land-
and seascapes.

In our red-hulled Folbot we were at least two dozen miles from the
nearest, emaciated trees, which resembled giant green pipe scrubbers,
on the other side of the Brooks Range. We were so far from radio contact that a communication device (other than a walkie-talkie to speak
to a plane overhead) would have been useless.[2]

We pitched the tent on a sandbar, several yards above the riverbank
piled with glacial-crushed and river-rounded rocks the size of lemons.
Dave laid the shotgun between us but left the safety on as we wrote
in our journals. We slept deeply and my journal (recorded in present
tense) tells about the next day:

Ranger Dave Buchanan drinking from the 59-degree water of Kugrak Springs in a
tributary of the Noatak River, 1983. Since the water doesn't freeze in winter, it hosts a
year-round population of American dippers. Dënéndeh, Kuuvuan KaNianiq, Gwich'in
Nàhn, and Iñupiat ancestral lands. JON WATERMAN

I jump up from deep slumber and reach for the tent door as Dave pumps a round into the chamber: Ka-chunk! We hear but can't see an animal run into the opposite side of river snapping willow branches. I unzip the door and shift back as Dave points the gun barrel out and the animal splashes and pants in heavy breaths right toward us.

We are immediately relieved to see that it's not a bear but a caribou that shakes off a shower of river water on the shore immediately below with her hooves throwing sand and she breezes past with the smell of wet sweaters and huge, brown darting eyes. All so close.

Jeez.

Dave says, "What's chasing it?"

Hooves shake the wet ground below us.

"It's a female," Dave says.

"How do you know?"

"Cows shed their antlers in spring and bulls keep 'em through the late fall."

Then another faint splashing as a large, blonde wolf swims the river and charges out of the water next to our kayak. Her muscles stand out in definition with wet fur plastered to her body. She lopes past utterly disinterested in us as Dave lowers the gun and we step outside to watch the chase fade in the citrus light of the early morning sun.

Back in the tent hoarfrost sifts, gathers, and drops off the walls, and my back is already wet from leaning on the nylon as I sit transfixed looking at the low light on the mountains. A bird I don't know calls huskily and Dave IDs it as a Bohemian waxwing while chickadees "chip, chip, chip" and the river swishes gently below.

I blow on my cold hands and sharpen a stubby pencil with my knife and search for the words to describe it all: Enjoying the reddening bearberry on the valley floor, the vast valley, IDing animal shit and finding chewed willows. The world here matters to me in a way I would never have guessed before this visit because the wolf might mirror my own life: her independence and the effortless manner of surviving in wilderness.

That's what I want too. Mastery.

An hour later, we pushed our tandem kayak into an eddy next to the wolf tracks. I put my palms next to her tracks on slightly dried mud above the wet riverbank in hopes that I could feel the power of her strides. Dave rolled his eyes with the knowledge that I'd get stuck in the mud. Sure enough, the floury till of rocks crushed in glaciers held me like quicksand and it was all I could do to muscle my feet back onto a rock.

Dave slid into the bow. Since my feet were already wet and muddy, I stepped onto a submerged rock to clean my sneakers in the water, and then pushed us into the current as I slipped onboard. With no head-wind, the current pulled us downstream and we steered with our paddles past a raven that croaked like a frog. The river had washed open its sides to reveal dirty gray permafrost hidden underneath, and we could feel the refrigeration of the ancient ice as we glided by the overhung riverbanks.

The ridgelines of the Brooks Range ran parallel to the river on either side of us and poked their contours up toward the sky, resembling the hulls of emerged submarines, frosted with early winter. The glacier ice on Oyukak Mountain glistened in the sunlight, while new-fallen snow on the lower mountains showed in vivid contrast against the blue sky. Streams ran in silver ribbons off the hillsides and carved gorges into the broad alluvial plains until they spat out a spray of boulders at their confluence with the Noatak.

Spooked caribou ran into the river with grunts and snorts as if chased by more wolves. They swam out into deep water and held their antlers

up with the aplomb and steadiness of waiters en route to the kitchen with full trays above their heads.

They were on the longest migration of any land animal on the planet. Many of the bulls and cows that passed us in the river had logged over two thousand circuitous miles in the last year across a roadless area the size of California.

While caribou-like figures were depicted in European cave art more than seventeen thousand years ago, the much older North American caribou originally came from South America nearly five million years ago, then spread north into Canada and Alaska. Pleistocene glaciers cut off and confined the woodland caribou to the forests and created a new and slightly smaller subspecies: the more northern, barren ground caribou of the Western Arctic Herd (the largest of Alaska's thirty-two different herds).

These herds define the Arctic. Over the tens of thousands of years since the ice melted and their populations grew, caribou have shaped the area's ecology.[3] The herd that sashayed in and around us on the river has long shaped subsistence and cultural practices for nearly forty communities in northwest Alaska. They call the caribou *tuktu*. Villagers have interacted with tuktu in the Arctic for tens of thousands of years.

Unlike the deer of the Lower 48, these are gregarious animals that run and swim shoulder to shoulder as synchronized dancers, with constant, minute adjustments of their bodies so they don't bump one another or clack antlers. They swam faster than we could paddle as their breath pumped vapor clouds into the air. When they aggregate into one massive crowd of thousands up in the summer calf grounds north of the Brooks Range, they effectively "swamp" and visually confuse hungry wolves and bears.

Even while absent from view, their sign is everywhere. Innumerable fecal pellets, dropped antlers, molded skulls, a leg bone here in the willows, half a rib cage there in the dwarf birch. Parallel crescent-moon tracks the size of my hands were stamped into most riverbanks.

3 While Iñupiat were sustained by caribou over many millennia, dropped caribou antlers are an invaluable source of calcium that sustains numerous small mammals—voles, mice, porcupine, etc.—throughout the Arctic. Caribou herds add nitrogen to relatively sterile tundra regions as they migrate north from the forests to the tundra every year. Since Arctic soil has very little nitrogen to support plants, the millions of caribou that urinate and defecate across the Arctic (with nitrogen-rich feces and urine) sustain what would otherwise be a plant-free barrens. Before the extinction of much more abundant Arctic megafauna thousands of years ago—such as the huge woolly mammoth and steppe bison—these nitrogen-spreaders undoubtedly shat the Arctic into a lush, plant-rich place.

Seen in its threadbare summer coat, as the Arctic warms, the red fox has begun to move north and displace the smaller, less dominant Arctic foxes. Iñupiat ancestral lands.
CHRIS KORBULIC

In rivers or lakes their hooves become flippers (while their body's guard hairs are hollow and buoyant in the water). By fall the callused parts of their hooves have been worn down by rocks and the soft pads have lost their hair as the hooves widen for more surface area across wet permafrost bogs. In winter, their hooves harden again on the outside and soften on the inside with a new growth of hair to insulate them from the cold. Then they use their concave hooves as shovels to dig through the snow for lichen.

The velvet skin that coats their antlers holds blood vessels that allow caribou to cool their body temperature in the heat of summer. Within weeks the bulls that clattered alongside us on the Noatak would shed their antlers. In spring, as daylight returned, their antlers would grow back at an inch a day.[4]

4 Caribou—like moose and deer—draw calcium from their bones to rebuild their antlers in less than four months. A bull caribou's antlers can weigh as much as thirty-five pounds.

All this animal activity amid such a primeval place put me into a new, instinctual level of awareness, along with the sensation that we had been transported into an earlier world. Like that of Lewis and Clark upon their Missouri, surrounded by a horizon of bison.

The tundra around us had reddened with frost. My toes felt numb inside neoprene socks. And I couldn't help but notice—as Dave had repeatedly pointed out—that only two days out and the polypropylene T-shirt beneath my fleece jacket emitted the soured, plastic-bag smell unique to 1980s synthetic clothing.

Then Dave whispered, "*Wolf*," and pointed to the left bank where a black wolf sat regally on its haunches, gone still as a statue, steadfast in its stare as if to critique our paddle strokes.

From my journal:

> *We paddle slowly for the shore and to our amazement, the wolf waits for us. So, we carefully look down at the ground and only steal peripheral glances his way as if eye contact will show too much dominance and violate the rules here in the high kingdom of animalia. As we step*

Mountain avens (*Dryas octopetala*) flower fossils allowed paleoecologists to identify past periods of climate-change warming called Older Dryas (14,000 years ago) and Younger Dryas (11,700 to 12,900 years ago). Schwatka Mountains behind, Alaska. Dënéndeh, Kuuvuan KaNianiq, and Iñupiat ancestral lands. CHRIS KORBULIC

out of the boat the wolf backs off into the willows, yet slowly, not really fleeing so much as inviting us to follow. So, we follow through chest-high willows. Each time I raise my camera the camera-shy wolf jumps behind another willow. Somehow, he is not afraid and carries himself with dignity. He ducks behind the willows two more times and each time I lower my camera until he reappears atop an old glacial esker out in the open.

He sidles along in a diagonal lope with his rear feet hitting the tundra to the side of his front feet, instead of moving straight ahead. Now that he's out in the open we can see—unlike the muscular blonde wolf of this morning—that this black wolf has a gimpy rear leg as if he'd been kicked by a caribou. His ribs poke out through gray fur. Still, there's spring to his steps.

We chase our conceit that this wolf will lead us somewhere important and sure enough on a nearby riverbank we hear mewing noises. Then the old gray-black wolf begins to lope south abandoning its diagonal sidle as if its mission is now accomplished. So, we push through waist-high willows for thirty yards and suddenly find ourselves in a clearing with the sandy wolf of the morning—who jumps down from her family at the den and sprints for the river.

"Jeez, Jon!"

"My god," I reply.

There are five pups lying in the sand next to the hole of their den. Most are asleep; one looks up at us from the leg bone of a caribou and tilts its head with curiosity at us then slowly stands and walks away while its siblings remain asleep.

We crawl closer; their fur riffles in the wind.

"Don't touch anything," Dave whispers. "Don't want to foul it with our scent."

At this the other pups yawn and stretch; one licks its paws. From the river below comes a howl from their mother, rising sharply, until it breaks into its third and final note after twenty seconds as she strains for more volume:

Owwwwwww—wwwwwwwwww-WOOOO

At this all five of the pups immediately stand up: twenty-five-pound tawny, cinnamon-colored furballs who stumble atop uncertain legs down off the den hill toward their howling mother. Rather than further violate their sanctity by chasing them, we take a photograph of the eight-foot-den hole and leave, still not speaking above a whisper. Like being in church.

We beat our way back to the kayak past bearberry and crowberry and mountain cranberry all brightly advertising their wares in an acerbic sweet fermentation. All the way we hear the seductive echo of the mother wolf's three-noted croon telling her pups to "come away, come away, come away."

How could the hair not rise up on the back of my neck?

We lift then shove the big kayak into the river, and slide into our cockpits with paddles behind our backs braced across the thwarts to prevent snapping the wooden crossbars. I cup a hand to drink from the river.

At the next bend we see all the pups swim the river with their muzzles held high and tails that work the current like rudders. We back-paddle to give them space and watch the pups climb out one by one. As the last skinny pup—maybe the runt of the litter—comes up out onto the silt bank a beat slower than his siblings, his legs are sucked down into the glacial till and he begins to whimper. He's trapped.

While the blonde mother sits still in the reddening tussocks above, refusing to look at us, we drift closer. As the trapped pup begins to whine, Dave and I look at one another with the same question on our

In the Noatak headwaters, the former tundra riverscape—recently invaded by tall willows
as the Greening of the Arctic transforms the land—is surrounded by lakes and wetlands
and mountains with their snow line a vertical mile above. Dënéndeh, Kuuvuan KaNianiq,
Gwich'in Nàhn, and Iñupiat ancestral lands. CHRIS KORBULIC

minds: Should we rescue it? Before we can paddle over to the bank, two other pups run down and bite into their brother's neck fur, then with three tremendous pulls—like a tug of war—they yank him out of the silt-like quicksand, growling and play-biting one another with their little needle teeth. The mother then leads all five off up onto the tundra bench. No one looks back at us and we watch only with side-long glances to respect their privacy.

That evening we made camp on an open sandbar. While we strolled a high bank downstream, we spotted a distant grizzly bear on the tundra. So with the stealth and deference for a creature above us on the food chain, we crawled on hands and knees up to an old glacial esker and laid down on top of the ridge a hundred yards from the grizzly. The binoculars revealed the details.

The grizzly sat atop the remains of what resembled a large bull caribou. The bear periodically lowered its bloodied muzzle into the bowels of the dead caribou and tore out its intestines.

Although the two animals were in ankle-to-knee-high dwarf birch and miniature tundra plants, the kill site looked like a bulldozer had cleared it to bare soil around the kill. You could just imagine the six hundred-pound, muscle-bound bear as it lifted the big caribou off the ground and flung it around to break its neck before disembowelment began.

On the sidelines sat a fox and a dozen ravens. At one point the fox crept in closer and the bear stood up and made a quick head feint—with all the adroitness of a point guard's head fake toward the net—and the fox darted away with tail between its legs as the birds rose in the air and settled back down again as if they had just ridden the crest of an ocean wave.

We, too, got the message and crept away. Dave asked why we had gotten so close to a grizzly on a kill, particularly with the shotgun back in the kayak. Grizzlies are the black holes of the Arctic universe, dominant force fields that absorb everything—leaves, grass, blueberries, roots, ground squirrels, or large-hooved animals—that enters their orbits.

At the same time, however, the presence of grizzlies—that need several hundred square miles to survive—showed us that the establishment of this park landscape had unequivocally succeeded.[5] Still, I had a lot to learn about grizzlies.

We reached the floatplane takeout at Lake Matcharak the next day, folded up our Folbot, hauled it a half mile to the lakeshore across a field of unstable tussocks, mercifully free of mosquitoes, and set up camp.

There were no jet contrails in the sky. And I couldn't get over the pressure of light against my face. I remember how I stared at it aglow on my hands.

The evening light flowed like viscous liquid into the tent and set the sedges afire as if the wind carried a molten wave. It painted whitecaps on the lake, colored the snowbound peaks pink, and lit clouds into brigantines of fire.

"You going to catch dinner?" Dave asked.

"I'll try," I replied.

The Alaskan myth of a fish on every cast comes close to the truth in the Arctic. In five casts, I caught two grayling. Compared to the more pensive salmon, grayling hit and run like underwater bandits and bend your rod double.

I cleaned the fish in the water with my Swiss Army knife and threw all but the filets out into the water where a bear wouldn't smell them. Then we sat fifty yards from the tent, torched the stove, and sautéed the firm white flesh—scales down—until it curled and cupped in the frypan. As we picked a few small bones out of our mouths and threw them into the lake, we knew we weren't the first to give thanks to grayling on this sandy shore.

As early as six thousand years ago—more than a millennia before the Egyptians began to build the pyramids—the ancients fished and hunted

5 Thirty-nine years later, Dave wrote to me: "My preference would have been to pack up and float downriver to put some distance between us and the bear. However, if it was mid-August, as I recall, the days would have been shorter to preclude that. Or maybe, we just enjoyed watching it and assumed the bear was well fed! I'm sure I didn't sleep well that night knowing we had a brown bear so close. Probably not the smartest move for a couple of rangers."

He added that on a river patrol three years later, his last, the wolf den had been abandoned. Dave believed that since the wolves were on the western edge of the park, they might've been killed by human hunters. Among his many trips into the park, our 1983 trip was his most memorable for the incredible wildlife encounters.

alongside the pure-blue, cloud-reflective waters of Lake Matcharak. Known to modern archaeologists as the Northern Archaic culture, followed by the Arctic Small Tool tradition (ASTt), named for the arrows and spearpoints minutely and carefully chipped from obsidian and chert, these hardy Iñupiat predecessors depended upon the caribou. The Canadian archaeologist Robert McGhee called them Palaeo-Eskimos, with no definitive link to the Inuit or Iñupiat (Eskimos) who succeeded them several thousand years later. Although their tools, hunt techniques, and houses mostly proved different from the more modern-day inhabitants of the Arctic, McGhee believed that they invented the igloo.

"The Palaeo-Eskimos provide an example of lives lived richly and joyfully amid dangers and insecurities that are beyond the imagination of the present world," wrote McGhee in his book *Ancient People of the Arctic*.

Park service workers dug out and sifted hundreds of thousands of animal bones and chipped-stone tools. Blades, burins, and scrapers were buried in the sand and frozen in the permafrost alongside the lake. Archaeologists have not found another Alaskan ASTt site that comes close to the number of ancient animal bones found at Matcharak. Most of the bones were caribou, hunted and butchered in summer, then eaten and used for clothes, skin-tent shelters, and tools. There were also the remains of Dall sheep and ground squirrel bones. Since no large wood was available, they had fastened their microblade chert and obsidian points into caribou-bone handles or spears.

Despite the hostile, cold environment and the lack of comfort or security in a hard world where the rivers and sea completely froze over, these caribou people designed caribou-bone tools with etched lines and artistic grace. Their carefully and slowly carved tools spoke of aesthetic sensibilities and a belief in a higher life beyond the brutal day-to-day realities of survival.

In little more than a couple of centuries, like-minded, artistically inclined hunter nomads from Western Alaska migrated and eventually

Ranger Dave Buchanan approaches an overhanging riverbank during a typically cold
August in the Noatak Headwaters, 1983. The North is a much warmer, different landscape
than existed forty years ago. JON WATERMAN

colonized all the Arctic. Given the iron-cold winters and short summers, it's hard to imagine the ingenuity and toughness it took for these people to prosper amid hard-to-catch animals and the frozen aridity of the North. They lived short lives.

They journeyed with their shaman leaders from Alaska more than two thousand miles east to Greenland in animal-skin boats. One of the earliest pieces of ASTt masks ever found—a 3,500-year-old, miniature mask portrait of these colonizers—was found in Eastern Canada. The tattooed face, McGhee said, "emanates tranquility and grace." Less than a thousand years later, Pre-Dorset people subsumed the old culture. They hunted marine mammals, developed the dogsled and seal-oil soapstone lamps, and continued to carve several-inch-long stone masks that depicted serene, half-animal, half-human faces.

Andrew Tremayne—one of the archaeologists who uncovered the remains at Matcharak and returned to the site for several summers—once held the tiny projectile points in his hand in wonderment. Since the earliest evidence of bow and arrow use belongs to the ASTt (and the tool likely came from Asia across the Bering Land Bridge), the projectile points could've been from arrows, or even spear darts from atlatls (spear throwers). He had seen stone tools from many time periods, but at Matcharak the ancients had taken the time and trouble to intricately flake even the smallest blades with a unique combination of artistic expression and tool function. Fascinated, Tremayne replicated the tools and found that even a tiny spall-flake proved incredibly durable and could cut or score animal bones.

"The level of craftsmanship on even their tiniest stone tools implies a great appreciation for the aesthetic value of the objects they produced," Tremayne said. "If their spiritual beliefs and stories are anything like the tools they made, this part of their culture would have been very intricate and beautiful. I often think they must have been the toughest people to have ever lived on Earth."

As would be true with the Iñupiat who followed in their footsteps.

Tremayne believes that the ASTt speared their grayling and may have created three-pronged tridents or weighted nets used by latter-day fishermen. Since the ancient inland peoples' tools were so tiny, the Iñupiat who now fish on the coast will ask Brooks Range travelers if they've seen any sign of the "little people."

Dave Buchanan and I went to sleep that night with grayling in our stomachs, beneath stars that winked and blinked light with all the phosphorescent sparkle of bioluminescent organisms in a blackened ocean. We looked forward to the next day's floatplane. We would be whisked back to the ranger station in Bettles, Alaska, into warm rooms with cold beer and a meal that we would heat in a microwave. Our casual patrol was a far cry from the hazard-filled journey that the ancients took in their caribou robes as they paddled and walked several hundred miles back to the sea thousands of years ago.

Barren Ground Grizzlies
1984

From 1946 to 1983, forty-eight park bears—that had gotten into human food or injured Denali National Park visitors—were relocated or euthanized. After an education program, along with mandatory use of bear barrels in 1984, backcountry incidents dropped by up to 90 percent. From 1983 to 2016, only four dangerous bears were euthanized, with three problem bears relocated (one to a wildlife park).

The barren ground grizzlies of the North are beasts apart. In the early 1980s my work as a ranger in Denali National Park allowed me to repeatedly cross paths with and interact with these unique bears.

We would find random grizzlies to test the early versions of bear-proof human-food barrels, which were smeared with fish oil to attract the bears. Out on the tundra I'd walk up to a bear and roll the barrel toward it, retreat, then watch to see if the bear could open the barrel to get the bacon inside. At that time, the park had the highest backcountry human-bear interaction rate in North America, but bear incidents declined after the foolproof bear barrels—dubbed Bear Resistant Food Containers—became de rigueur for backcountry campers.[1]

These three hundred- to six hundred-pound barren ground grizzlies are distinctly different from the salmon-fattened Kodiak or brown bears of southern Alaska that can weigh up to 1,500 pounds. A scarcity of food in the Arctic makes the barren ground grizzly smaller, more aggressive, and likely to range over a thousand square miles to find food, while the well-fed southern grizzlies live within a dozen square miles of plentiful salmon streams. The regionally separated bears are as behaviorally different as a hungry rottweiler and a well-fed golden retriever.

PREVIOUS SPREAD: A subadult male grizzly (brown bear) in Denali National Park stares at the camera. Several hundred grizzlies roam the north side of the park, among an Alaskan population one hundred times larger. Upper Kuskokwim and Dënéndeh ancestral lands. JON WATERMAN

I was the only ranger imprudent enough to sign up for bear-aversion therapy work, which, I soon learned, would effectively make me grizzly bait. First, we would locate the radio-collared grizzlies that had harassed backpackers or eaten their food. I would then set up a tent and entice the grizzlies into camp with the distinctively scented pork ramen soup that, once boiled, carried to the bears' noses on the wind. Then the excitement began.

While these bears' brains weigh a third that of a human brain, their olfactory bulb is five times larger than the pencil-eraser-sized olfactory bulb of human backpackers. Add to this millions of scent-sensitive nerves housed inside their huge snouts, and they can smell at least two thousand times better than me—let alone any land animal on the continent. They could pick up scents from out of sight, even miles away, particularly with a favorable wind.

Through binoculars, I could see the bears lift their heads as the airborne soup molecules tripped their olfactory apparatus, but then, invariably—as if they sensed my presence—the bears feigned disinterest as they ambled slowly and circuitously into my camp. As soon as they came within twenty yards of the tent, where I hid with a shotgun, I aimed for their fat-swaddled hindquarters and blasted them with a rubber bullet. The bears always sprinted away, hugely chastened.

Aversion therapy to teach bears to avoid backcountry camps proved more effective than the usual method of a tranquilizer dart and transport to a remote corner of the park.[2] If bears couldn't be trained to avoid people, someone might get hurt or killed and then bears would get euthanized.[3]

Bears often scratch their backs like cats (on a boulder), graze as constantly as cows (at a blueberry bush they'll rake whole branches through a gap in their molars and strip off the berries), and can run as fast as horses (as they sprint toward a ground squirrel that ducks into its hole). When they stop to dig up a ground squirrel—and fling boulders backward through the air—their strength is unmistakable. As

2 One grizzly helicoptered around the south side of the mountain to sea level walked over a ten thousand-foot pass and through a hundred miles of ruggedly crevassed glaciers to return to its old territory of backpackers on the north side.

3 In 2012, a lone forty-nine-year-old backpacker (unarmed and without bear spray) stopped to take over two-dozen photographs of a grizzly that grazed in the brush less than fifty yards away—the park recommends that backcountry visitors keep a quarter-mile distance from bears. The last picture in the camera showed the six hundred-pound bear approach the man, who was killed, dragged into the brush, and buried in a cache where the bear partially consumed his body. Wildlife officials shot and killed the bear a day after the body had been discovered. It was the only park grizzly-human fatality in more than a century.

they stand up on their hind legs to get a better view, they have an eerie resemblance to humans—we share up to 90 percent of the same genes. (Inuit stories tell of grizzlies who remove their furry skin to reveal humans who hide inside.) So, when a grizzly comes to investigate you, the temptation to turn and run like a prey species can be hard to resist.

Aside from those few encounters when I shaped bear behavior with rubber bullets, I mostly got schooled by grizzlies on backcountry trips. I learned that it is essential to manage your fear when you meet a grizzly in the wilderness.

Once, on a three-day, solo patrol across the Alaska Range, a bear surprised me and popped out of the brush fifty yards away when I stopped to disassemble and clean my handgun on a T-shirt. As the bear walked toward me, I shouldered my pack and, with an act of calm, slow movements, crabbed sideways up a tundra hill with the T-shirt and gun parts cradled in my hands. Without direct eye contact, I gazed sideways at the blond bear that outweighed me by a couple hundred pounds, and I talked out loud about how nice the weather had been (to show that I was not the usual prey species). Once on the hilltop, twenty feet higher than the bear, my height advantage deterred him. As Blondie ambled off, I quickly reassembled the gun. While unnerved by the encounter, it seemed clear that the bear was merely curious.

I also learned that bears demand respect. On other occasions, when forced to retreat, I did so slowly with a muscle-bound gait as if I, too, were a species not to be trifled with. I never ran. I avoided fidgety, prey-like movements. I repeatedly took to higher ground or even stood atop boulders to appear larger than the bears I met. And I never underestimated their vision, particularly at night, when they could see better than me.

I learned to talk to bears in a deep voice to reassure them and at the same time show deference, without aggression. It also proved essential to act confidently near bears because they can smell and sense fear.

My younger self on duty during a patrol—supplementing our freeze-dried food diet with freshly caught silver salmon—in the Tokositna River headwaters on the south side of Denali National Park, 1984. Dënéndeh and Dena'ina Ełnena ancestral lands. JON WATERMAN COLLECTION

Petals closed in the rain, the heliotropic Arctic poppy will open and rotate with the sun, its petals reflecting light to the pistil. The cup shape traps light and air inside to warm the ovaries. Iñupiat ancestral lands. JON WATERMAN

In the Arctic I abstained from soap, deodorant, and minty toothpaste, and I kept all my food outside the tent. I usually cooked away from camp and stacked pots on top of one another—so they would fall over with a clang and wake me if a bear wandered in—atop the food bear sacks or barrels.

Grizzly encounters were always a highlight of my journeys in the North. As a keystone species, bears keep animal populations in check, disperse nutrients throughout the ecosystem, and, like no other animals I'd met in the wild, induce humility and show humans our place in the food chain. Spend enough time around the barren ground grizzlies and you begin to recognize a remnant of ourselves in the evolutionary cycle: they play, show curiosity, stand on two feet, and omnivorously eat the same foods as we do. Without grizzlies, the Arctic wilderness would be incomplete.

A Qallunaat's Education
1997

Trip Plan: *More than a dozen years and Arctic trips later, in early July I drove from Colorado in my truck for three days as far north as I could go into the Northwest Territories. From the town of Inuvik (population 3,361), I flew in a small plane thirty-five miles southwest across the vast bogs of the Mackenzie River Delta to the village of Aklavik (population 739). From the westernmost channel in the labyrinthian delta—more like a tilted swamp than a river—my plan was to reach the coast sixty miles downstream and solo paddle west along the Beaufort Sea coast to Prudhoe Bay, Alaska—more than four hundred miles in total. Over the next couple of years, my dream was to cross the roof of North America along the Northwest Passage.*

Clear of the muddy, mosquito-infested delta on the afternoon of July 20, I emerged from a tangle of whitecaps in my kayak and high-stepped out onto the pebbled sand of Shingle Point, a peninsula that kicked out into the Beaufort Sea like a mile-long caribou leg. The Canadian Inuit camp there included several dozen white A-frame tents and plywood shacks.

PREVIOUS SPREAD: Muskoxen protectively surround a calf on an island in the Beaufort Sea. As the Arctic warms nearly four times faster than the rest of the world, rain-on-snow events imperil this amazing Ice Age species. JON WATERMAN

A fly-hung husky with matted fur barked and jumped against its chain. Smoke blanched out of tent stovepipes. The tang of fish hung in the air.

I hadn't seen or spoken to another person for three days since I had left the Mackenzie River Delta and felt uneasy because I had heard (from a surly white bartender in Inuvik, Northwest Territories) that the Shingle Point whale camp was off-limits to white, tree-squeezer environmentalists like me. So, I couldn't have been more pleased when a thirty-something hunter with wide cheekbones and canted eyes—surrounded by a gaggle of children—walked over to me, held out his hand, and said, "Welcome to Shingle Point."

"Holy boy!" one of the kids shouted while he pointed at my Klepper. "What kind of a boat is that?"

I explained that it was a kayak invented by their ancestors. The boy, Fred, gobsmacked, watched my eyes and listened carefully as I showed him how to feather the kayak paddle with a wrist twist before he arced it into the water. After the lesson, I put him in charge, with a plea to take special care of the boat.

He had large brown inquisitive eyes and a burn scar on the left side of his face. I held up a hand and we high-fived. "Thank you, sir," he said.

With Fred's tutelage, the boys and girls (some of whom were still toddlers) spent the next several hours—until they turned blue—as they took turns in the boat out in the calm lagoon, while the nonpaddlers stripped off their shirts and raced behind the kayak and swam like jubilant seal pups in the frigid water. The adults retreated to a cigarette-smoke-filled shack out of the icy wind, where they served me coffee and fish eggs.

After repeated trips to the Noatak and the wilderness throughout the Brooks Range, I had come here to interact with the people of the North. To really learn about the Arctic, I figured, best to hang out with those who lived here. Across the Arctic, they are known as Inuit (the people, who seldom identify themselves as Eskimos—eaters of raw meat, as the Cree reputedly called them). Alaskan Inuit on the northern coasts are Iñupiat (the real people, and some still proudly refer to themselves as Eskimos). Aside from similar DNA, northern Canadian and American Indigenous people are often related—and unlike the polyglot of Lower 48 non-native Americans—most could trace their DNA to Siberians.

The people called me *Qallunaat*, defined as those who are not of the North. But rather than a skin-color reference, to be a Qallunaat has more to do with southern peoples' habits: the suppression of belches and farts, the strange salutations, and the way people like me have colonized the world—how we have changed Earth's temperature and caused air pollution that wafts north and drops PCBs into Arctic waters, absorbed by the seals and whales that the people eat. Still, these definitions of Qallunaat are more wry observations than critical judgments.

Peter, the elder in the shack, said they used to dip frozen whale into seal oil. "But now we no longer do this because the blubber in seals and whales is polluted." He confessed that he personally still ate the blubber, then joked that he liked how it made him glow in the dark.

Several of the people I sipped coffee with were from Kaktovik, Alaska. After a visit with Canadian neighbors and family, they would make the long journey home in their open skiff with an outboard motor back across the unmanned border. By my arrival alone in a kayak—the mythical tool of their ancestors—I would make fast friends everywhere in the Arctic. (Still, even if I flew in like most tourists, these generous people would have made me feel welcome—albeit without the same inquisitiveness.)

If I could tolerate the alone time necessary to paddle four hundred miles west from the Mackenzie River to Prudhoe Bay, my plan was to paddle east over the next two years across the Northwest Passage to continue my Arctic education. My curiosity—first stoked by the 1983 Noatak trip—hung on the wonder of the landscape alongside the Arctic Ocean (which encompasses the Chukchi and Beaufort Seas), where the

In 1997, as part of my journey across the Northwest Passage, I paddled alone over four hundred miles from the village of Aklavik, Yukon Territory to Prudhoe Bay, Alaska—a very isolated stretch of coast.

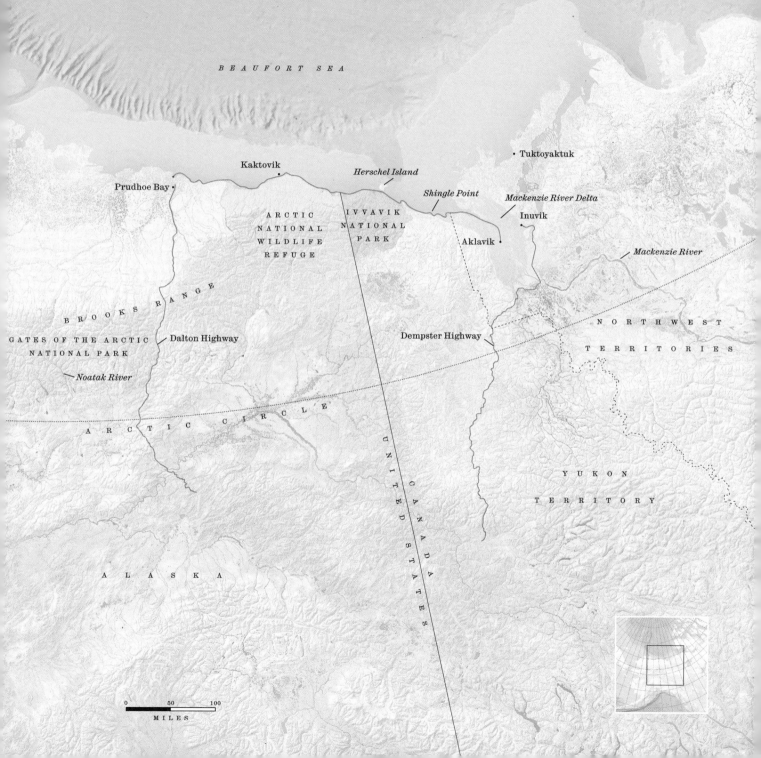

BANKS
ISLAND

Sachs Harbour

BEAUFORT SEA

Tuktoyaktuk

Kaktovik

Herschel Island

Prudhoe Bay

Shingle Point

Mackenzie River Delta

ARCTIC
NATIONAL
WILDLIFE
REFUGE

IVVAVIK
NATIONAL
PARK

Inuvik

Aklavik

Mackenzie River

BROOKS RANGE

GATES OF THE ARCTIC
NATIONAL PARK

Dalton Highway

NORTHWEST

TERRITORIES

Dempster Highway

Noatak River

ARCTIC CIRCLE

YUKON

TERRITORY

UNITED STATES

CANADA

ALASKA

0   50   100

MILES

eerie penumbral light and abundant wildlife lit my imagination and repeatedly appeared to me in my dreams.

But in what had already become the nightmare of the climate crisis, by 1997 scientists had shown that the Arctic was warmer in the twentieth century than at any time since historical records began in 1600. Satellite photographs from 1997 showed that the summer sea ice had shrunk by half a million square miles since 1979. (From 1979 to 2022, sea ice has declined by 40 percent.)

At the time of my trip, feedback loops in the Arctic were not widely understood. An amplified feedback loop causes accelerated changes; a stabilized feedback loop (such as cloud cover that allows the Arctic to remain cool) decreases the effects on the system. A century ago, the white surface of sea ice naturally reflected heat back into the atmosphere and kept the Arctic cold. But an amplified feedback loop began in the mid-twentieth century as the loss of sea ice exposed dark ocean waters that absorb rather than reflect heat. This loop causes an increase in temperature and further ice loss on both the sea and the land.

Until my 1997 visit, Shingle Point summer temperatures had risen by 3.6 degrees Fahrenheit over forty years. As the sea ice continued to melt, many of the Inuit believed that their waters would soon be navigable by commercial ships all summer long without icebreakers. Ships that could travel from Europe to Asia through the Northwest Passage would save several thousand miles and weeks of travel compared to the Panama Canal or around South America. Yet the potential of ship animal strikes, increased pollution, and a noisier ocean would harm many vulnerable animal species.[1] Loss of wildlife—that, from the Inuit perspective, equals food security in the Arctic—would, in turn, have a huge cultural impact.

In the Shingle Point shack, the elderly Peter—who lived 350 miles away on Banks Island—talked about how there were no mosquitoes in the 1960s, but warm temperatures had now brought them to his island

1 In 1997, a total of two medium-sized ships—an icebreaker and a cruise ship (which I crossed paths with after I left Shingle Point)—made it through a less icy Northwest Passage. In 2022, twenty-six ships made the passage; eight were large cargo ships, only one was an icebreaker.

In strong winds and icy water, I often reveled in my sense of self-sufficiency in such a remote and beautiful seascape. When the wind blew, I attached a higher mast section and sailed. JON WATERMAN

town, Sachs Harbour. Or about how they used to run sled dogs in early July, but now snow is gone at that time of year. He talked of species—bluebirds, robins, red salmon—he had never seen before so far north.[2, 3]

With a sudden burst of static over the CB radio in the shack, an Inuit hunter blurted out, "Coming in with beluga. Beluga!" It would be their first whale of the summer.

From the Noatak River Delta to Greenland, the people have depended upon and hunted whales for thousands of years. Armed with harpoons in large skin boats or in one-person kayaks (and now with rifles in motorboats), the people have harvested whales long before ranchers in the West had begun to raise beef. While the people had never reduced animal populations, by the late 1800s commercial whalers from the south had decimated bowhead and beluga populations from basecamps on Shingle Point and nearby Herschel Island. At the same time, whalers and Inuit hunters slaughtered thousands of musk oxen. The shaggy animals were eliminated in much of Canada and extirpated in Alaska. DNA studies have shown that the change in climate since the last Ice Age also played a part in their demise.

We left the shack and joined several dozen more Inuit gathered on the beach. As the boat approached with the whale towed behind, the people cheered, raised their arms, and bustled with excitement. As soon as the aluminum skiff ground its hull against the shore, a hush fell over the crowd as a half dozen people dragged the ten-foot-long whale by its tail fluke out of the water and up onto the sand. Instead of a dorsal fin, a slightly raised ridge ran down its back and I couldn't help but stare at what resembled Botoxed lips opened to reveal teeth that could have been a yellow zipper on a ghastly white purse. Its motionless eyes lay directly behind, and there was the melon-shaped sonar dome above its forehead.

No one spoke. It was as if we all bore witness to a dead human rather than a whale carcass.

2   By 2006, abnormally warm temperatures allowed grizzly bears to expand their range to Banks Island where they met polar bears. That year, a hunter from Idaho shot what appeared to be a tawny polar bear with a strange hump of muscle on its shoulders that looked like it belonged to a grizzly. DNA tests confirmed that it was the world's first wild grizzly-polar bear hybrid. Since then, a couple of generations of the hybrid bears have been confirmed in the Canadian Arctic Archipelago.

3   In 1951 Rachel Carson wrote in her book *The Sea Around Us* how southern biosphere birds such as Baltimore orioles, greater yellowlegs, and the avocet had spread into the Arctic in the first half of the twentieth century because "[t]he frigid top of the world is very clearly warming up." But she attributed it to a shift in temperatures caused by the tides, as per an obscure theory of the time—the world had not yet accepted that the climate had changed due to human emissions of greenhouse gases.

Two women stepped into the silence with their ulu knives and sliced off the prized fins and handed them to young Fred, who had pulled my kayak ashore to help with the whale and had pocketed his knife and a small piece of wood he had carved. He lifted the beluga fins onto a driftwood rack high above the reach of several dogs.

The men smoked cigarettes and watched as the women swung their arms and made precise cuts in the whale with their *uluit* (plural of *ulu*), without talk, as if they'd done it many times before. Across the North, the people believe that an ancestor's knowledge is held in the curved knife and passes on to the next generation of women who wield the ulu. The tool—one of a few used only by women in their culture—is believed to be 4,500 years old and cuts as efficiently as the micro-bladed points of the ASTt. Unlike a conventional knife blade, the arced curve of the ulu transmits a dynamic cut to the middle of the blade and allows fast, effortless one-handed strokes to slice food.

As the whale's purple guts slithered out and snaked across the Mackenzie River–browned sand, one of the younger boys splashed in blood-red water that surrounded the whale and shouted, "Holy cow!"

It took a couple of hours to strip out several dozen two-foot squares of white-skinned, orange-on-the-inside blubber (muktuk) and black meat to share with the camp. As one woman further opened the lower innards with a long slice down into the belly, an elderly man reached into the cow's uterus to pull out a pink, hand-sized embryo—its flippers like miniature hands—for the children to see.

The elder cradled the fetal beluga with solemnity. The man had grown up in a time when his parents had taught him that the animals they depended upon shared the same souls as people. When killed, the man believed that animals' souls departed their bodies and if the people showed disrespect or wasted any part of the animal, its soul would not pass on to a future life and would then haunt the irreverent hunters.

The man shook his head at my camera and asked me not to take a picture, even though I had asked and received permission to do so earlier. So, I put down the camera. The man smiled at me, toothless, sad, as if to acknowledge that life is difficult and cruel, but that I would understand that he worshipped both the dead whale and the unborn whale that he held carefully with both hands.

Later, as I prepared to continue my paddle east toward the Alaskan border over one hundred miles away, Fred handed me a three-inch-long piece of whittled-down, oblong-shaped driftwood. This was a practice done for centuries by his people, and by the previous Dorset culture of stone huts, and in the olden-day nomadic camps of the Arctic Small Tool tradition. I turned the driftwood over in my hands to discover a replica of my beamy kayak, carved with an upswept bow, cockpit hole, and rudder.

"Good skill on your journey," he said.

—————————

Days passed as I gave myself to what seemed a dive into the unknown. My Arctic apprenticeship deepened.

Along the littoral, the chestnut-backed, black-and-white-bibbed ruddy turnstones pecked at tiny bits of krill, then flew off on wings that fluttered as fast as butterflies. Globular pebbles gleamed on the beach and darkened with waves, then gleamed and darkened again, and gleamed until I looked away, half hypnotized.

I passed icebergs that initially appeared as giant birds, a whale tail, and an alligator in repose. In the Arctic, inanimate forms that resemble animals abound—rocks on the shore, driftwood, and clouds—and they seem animated. I wanted to believe it wasn't just my imagination.

At first the well-lit nights held a pale moon that floated freely on its own, a celestial ambassador for barely visible stars. I often paddled through midnight calms, enveloped in an air of tranquility that I wore and took comfort in like a blanket.

Inuit women butcher a beluga whale with ulus—used for several thousand years—on Shingle Point, Yukon Territory, 1997. The people depend upon sea mammals for food security and have not eradicated any creatures of the North. Iñupiat and Inuvialuit ancestral lands. JON WATERMAN

91

Belugas passed below as wavery ghosts; when they surfaced, they startled me as they blew and we shared air together. I had to take deep breaths each time it happened. I envied them. I wished I could swim with the same grace that held them snug as Lycra. I watched them as they spun and twisted untethered by gravity, free as hawks on the updrafts. *How must it feel to be so unencumbered?* I wondered. If I fell in my world, the air would not sustain me.

Eiders ran into the water from the shore, their grass-colored, white-patched heads flashed as they swam into the air. Never did they fly over land.

At the Alaskan border I paddled past the miniature Eiffel Tower-shaped, iron radar beacon, guarded by two customs officials who stirred—heads cocked and wings a-wiggle—as I greeted them aloud. But the ravens flew off without a reply as their black feathers creakily snatched the air. I closed my eyes and listened until it went silent again.

The distant ice pack fell and then bellowed up in accordion-enlivened mirages out in the infinite plain of the Beaufort Sea. I passed thousands of flightless old squaw ducks in molt—lined out in mile-long rafts—as they gabbled turkey calls, hiccupped with strange gasps, and dove under my bow. No doubt they wished to get back in the air as badly as the geese in molt along the shore: all a-sprint in awkward squadrons on their webbed feet like toddlers that had just learned to walk.

After a week alone, without sight of another human, I had talked to every animal I encountered, but fortunately, no one talked back. Whenever I felt weary, I took out the huge bag of dried whitefish that Peter had given me at Shingle Point. The two pounds of dried fish fat and protein fueled me with several times the energy of any candy bar. "Inuit power bars," he had called the bag of caloric energy.

I was alone, unheeded, and no one—not even my parents—knew where I was. I was forty-one years old with the endurance of a young man and the heart of an adolescent. I only considered my mortality while on risk-filled expeditions (otherwise I could not have contemplated such a potentially hazardous solo journey in the remote North).

I didn't carry a sat phone. And I deliberately didn't bring a gun, but it didn't matter because both grizzlies I paddled up to bolted away across the tundra as soon as I said, "Hello, Mr. Bear!" Most of the time I felt confident; although I occasionally felt awash in emotions and loneliness—the utter weight of unloved solitude. I thought it would be wise to wait and hold sentimentality in check until I reached Prudhoe Bay. So, I stuffed my anxiety away. It would be all too easy to let go and lose my center in this vast wilderness.

While this solo venture might have been one of the hardest things I had ever done, I also felt incredibly curious every time I turned a new corner of coast. I bowed to the sunsets. And I became attuned to the Arctic world around me in an instinctual way. How the sun's rays swept in long horizontal arcs through the dawns. The distant hum that emanated from the sea on quiet days. And I looked forward to wildlife encounters or new horizons so eagerly that I all but leaped out of my sleeping bag each morning.

I didn't bring alcohol. Once a day I rolled a single cigarette (even though I didn't habitually smoke). Cigarette fumes quelled the mosquitoes and provided a daily break from the monotony of paddle strokes. The nicotine and tar ramped up my senses as I sat far from shore and contemplated the rise and fall of the ice pack and the sea that rolled beneath my damp rear end and searched the horizon for the telltale spume of whales that blew mist into the air.

I listened to the steady thrum of surf on the shore, the yawp and purr and yodel of distant birds. And again and again, I listened to that inexplicable hum—like the vibration felt under power lines—that came out of the sea to the north. For days I felt flummoxed and intrigued by the noise, but then I gave up on an explanation and surrendered to yet another mystery of the North.

I stowed my watch and used the passage of the sun to mark time. If tempted to feel bored with the thousands of paddle strokes it took to make progress, I sang. Although agnostic, I repeatedly sang Henry van Dyke's "Hymn of Joy" (to the melody of Beethoven's "Ode to Joy," and substituted, at least in my thoughts, the song's references to God with the Nature that surrounded me). I belted out Tom Paxton love ballads; I sang the Kinks' "Apeman." I repeatedly revisited Joni Mitchell's "Urge for Going."

I relished the ease and comfort I took from a sound sleep on the ground—alert to noises around me. One night a cross fox disturbed me—I didn't hear him so much as I felt his presence outside the tent. I zipped open the door and he replied with a shake of his matted, brindled fur, licked his nose, and while I carefully avoided a direct look into his eyes, he padded away with an indignant rearward glance, quiet as a burglar. A skunky pong remained.

Some nights I dispensed with the tent. I didn't usually get cold and even felt tropically warm under my neoprene surf shirt.

Water is plentiful throughout the Arctic, but you have to hunt for it carefully. I didn't carry a filter or more water than what my liter bottle held, and at night I used my stove to melt water from bergy bits that littered the shores like glass blocks. When there was no sea ice to gather, I walked out onto the tundra and filled my pot and bottle from shallow puddles of tannin-colored water that had melted from the active layer of permafrost. But I avoided pockets of microbe-rich water with oily rainbow sheens, and I abstained from large or animal-tracked streams that could have been contaminated by giardiasis. Sometimes I put a few drops in my mouth to make sure the water didn't taste metallic or bitter. If I found bear sign—fresh tracks, scat piles, or roots that had been dug up in the tundra—I would push my kayak back into the sea and search for a safer camp.

When mosquitoes arrived in thick clouds and broke into shadow fragments funneled in the air around me, I paddled farther out from shore.

After innumerable blood draws, I developed a tolerance and the bites no longer swelled up or itched. Instead of a misery, the mosquitoes became a mere annoyance, a simple price to pay in order to strive toward Arctic enlightenment. Besides, without protein-rich mosquitoes as a food source, there would be a lot fewer birds. And without mosquitoes—more prolific plant pollinators than the birds—there would be fewer flowers.

On surfy days, I avoided wet rides and waited for sand spits or protected bays where I could duck around the waves into calm water. When I couldn't avoid the waves, I surfed to shore joyfully. It didn't seem foolhardy, because I couldn't have taken a solo trip across the Arctic if I were risk-averse.

I had always yearned to be immersed in wilderness, on a journey of self-sufficiency. I wanted to take a final exam for all my expeditions. I wanted to test strong yet vulnerable as I stood in awe of the landscape and its animals. This meant I had to guzzle large doses of anxiety. Anxiety about loneliness, or bears, or an early winter that could trap me on the coast. Or anxiety about how I would hitchhike all the way back to my truck in Canada (I didn't have the money to fly). Invariably, anxieties ballooned into fears that robbed me of sleep—and the list grew.

Hypothermia.

A current that would pull me out to sea.

Darkness—when I wouldn't be able to see animals outside my tent—as August approached and the sun began to set again.

A capsize offshore (it would've been impossible to roll my beamy kayak and I didn't have a dry suit).

And always, *loneliness*, but this diminished each day as I grew more accustomed to alone time, time away from people. Oddly, the enjoyment of solitude conferred a strength and power that I hadn't felt before. I didn't have to consult a partner about whether it was safe to paddle each day, or when to stop, or what time we should get up in the morning. Solo travel in the wilderness also breeds great efficiency and speed; it made me strong. Or so I repeatedly reminded myself.

Like the mosquitoes, daily anxiety and even fear seemed part of a necessary rite of passage in my quest to understand the Arctic. My fears about the ultimate screwup—death—effectively supplanted the silly worries about my impoverished career as a writer. Or the need to make money (even though I had no debt, aside from a six-month storage rental, and only owned a truck). In the workaday world, I would have become disenchanted, bored, or listless as I was forced to confront the nine-to-five existential qualms. Out here, with survival fears, the thought of a therapist seemed silly and self-absorbed.

Fear can release a certain power. Not the average, pseudo fears that we all process in our domestic lives, but the nightmarish fear that some ravenous quadruped could kill me as I slept and then gorge on my entrails. Or that appendicitis might poison me with toxins before I reached the next village. Fear of death. I figured that if I could vanquish these fears and their anxiety catalysts, then I would return to the normal world a humbler man. And once there, anything—marriage, children, a respectable career that did some good for the world—would be possible.

To quell moments of doubt, I stayed busy and pretended to own mastery—and even power—while I looked for animals, counted birds, contemplated icebergs, or sought driftwood for a shelter. Driftwood abounded on the shores, but I eschewed campfires (which would attract grizzlies or polar bears) and instead stacked the wood around my tent for protection against the wind. Campfires, after all, were for those who didn't hydrate, eat right, or dress properly. I didn't need campfire light for cheer or warmth because the Arctic is already suffused with warm, cheerful light. When I got cold, I climbed into my sleeping bag or went for a long walk across the tundra.

While I talked aloud, I also wondered if too much solitude would subvert my need for friends and family, or the persistent call of home. If I

To bear witness to the wonders of the Arctic, I had to learn to withstand the onslaught of mosquitoes, which could be avoided in the wind, far offshore, or in my tent. JON WATERMAN

would be tempted into misanthropy. But this was mastery perverted, which unnerved me, so I remained committed to my original goals: to find mastery in solo and self-sufficient wilderness travel, to learn about the people of the North when I resupplied in their villages or camps (along with my plan to live in two villages in winter), to meet a polar bear, and to carefully carry my journals and knowledge of the Arctic home so that I could share it all with others.

Mostly I looked and listened, and I gave myself to the land and sea as if to develop a sixth sense (and did my best to avoid the treacly trap of self-absorption). Yet if the past week of solitude had been hard, it would be mere training wheels for a continued journey east across the Northwest Passage, which would demand multiple weeks alone.

After eight days of solitary nonconfinement, two Iñupiat boats from Kaktovik loaded to the waterline with caribou meat pulled up (before I could see them, I heard the strange out-of-place putter of Evinrudes). As they drew close, they shut down their engines. Our boats rose up and down on the cold swell. The ice pack lined the northern horizon like a portent of endless winter. One of the elders smiled toothlessly at me and I heard myself say—dumb and nervous about a reciprocal conversation—"Hello."

The elder replied, with a lovely inflection on his end vowels, "We thought you were a piece of driftwood."

That instantly made me feel better—I had not slipped a gear—I wasn't the only one who perceived animate objects as inanimate in the Arctic.

Thomas, the elder, offered to feed me and then gave me brief directions to their camp: "where the land curves around again." (I loved this word picture and when I reached the far point, it looked exactly as Thomas— who shunned maps—had described.) They said nothing else, and I felt grateful for Iñupiat reticence because I didn't know if I could hold up my end of a real conversation.

As they motored off, I burst into tears.

That night when I caught up to them in their cigarette-smoke-filled tent, they fed me the *tuktu* they'd shot earlier that day, and although I didn't normally eat red meat, I ate, thankful for the hunters and the animal that had given its life for us.

Thomas, who did not eat, said that three dozen polar bears had come into Kaktovik recently after a whale kill. This proved his segue to ask if I had seen the polar bear ten miles back. I hadn't.

Thomas replied with a smile, "He was following you."

I didn't doubt it. After a long silence, Thomas asked what kind of gun I carried.

"None," I replied. Which sent silence into the tent for a minute.

Thomas, ever polite, then asked, "What you do if you meet a polar bear?"

And just as I had done with Peter at Shingle Point (also concerned for my safety), I reached into my pocket and showed Thomas my Sharpie-sized, Canadian "bear-banger," which shot a spring-loaded flare into the air. I didn't mention that I had shot it in the air above two grizzlies along the shore in Ivvavik National Park and they didn't budge after it exploded over their heads. Thomas looked unimpressed and took a long drag from his Marlboro.

As if to reveal my hand, I pulled out the air horn and placed it on the piece of plywood atop driftwood logs that was their card table. Thomas remained poker-faced.

So, I pulled out my can of bear spray, and with a flourish, tapped it down on the plywood with a smile, as if this, at last, would earn their respect. Smiles crept onto the faces of the five listeners, who all looked to the elder, culturally designated spokesperson who would educate strangers.

Thomas did not smile. "That," he said, and lifted a shaky index finger toward my can of bear mace, "just piss the polar bear off!"

In the village of Kaktovik on Barter Island, the Iñupiat woman at the food co-op gave me a bag of canned food but refused my money. Word had spread, through the hunters who had fed me, that *Qallunaat* ate only freeze-dried food and traveled alone in a kayak without a gun. I booked a night in the decrepit, diesel-fuel-stink, generator-drone of the Waldo Arms Hotel. I slept poorly and wished that I had pitched my tent outside. While no one wished me good luck as I left the next morning, two different Iñupiat wished me good skill—which I thanked them for.

I paddled angrily alongside the Arctic National Wildlife Refuge—a long-sought-after quarry for oilmen who wait to exploit the coastal plain. While Canada deferred to the rights of both Indigenous people and wildlife (through the Inuvialuit Final Agreement), and closed development in the adjacent Ivvavik National Park, United States politicians can't decide between oil or wilderness. From previous visits to Prudhoe Bay, I knew it would be impossible to have both.

All that day, the north wind pushed the ice pack closer to shore. Winter opened its jaws as a moody wind tore off the North Pole and blew raw against my reddened face. Snow winged sideways. The sea rose and bumped me up and down like the carnival horse my father had once lifted me onto. Like that day when I held tight to the plastic horse-head handles, my hands stayed warm in the pile-lined pogies against the fiberglass paddle shaft.

I felt engaged with every paddle stroke. I could feel the power travel up my legs, into my torso, and up my arms as I sent strength into the paddle and leaned with the waves and rode the water with the knowledge that nothing could stop me from the beauty of my dreams, from my quest for wonder. I wouldn't let the cold or the wind faze me. This was mastery. Power. To understand the Arctic, I figured as I pulled ashore, it would only be fair to take risks, to suffer.

I pitched the tent and blew on my fingers, then shoved them in my crotch as I jumped up and down to stay warm. I dove in out of the wind with gratitude. All night long, the nylon walls snapped and pumped and billowed with the wind.

In the morning, snow continued to fall and since the sea rose up in a white rage, I spent the day with a Patrick O'Brian novel in the tent. To short-circuit my fear about entrapment on the coast, I pulled out my maps and plotted how I would walk out if the sea ice blew in. Maps always gave me solace. Meanwhile, the nights had turned dark, but with the storm clouds, there would be no chance to see the northern lights. At least I could listen for them.

On August 11, as winter seemed to relent, I paddled out to Flaxman Island to investigate what appeared to be a row of rusted oil barrels. But instincts now fully primed from all the alone time—attuned to the wild world that cocooned me—the reptilian, predacious part of my brain sensed something entirely different: maybe a flash of light from a set of eyes, a flinched movement, or color that did not fit the usual inanimate objects. I kept my head down until, after an hour's paddle across the wind brought me close enough to look up without the binoculars, I could see six musk oxen. With their fur draped to the ground, their legs were hidden under their robes, and as they stood motionless to conserve energy, the animate had once again morphed into the inanimate.

I paddled to the opposite side of the island where the musk ox wouldn't see me and crawled up a hill, downwind—past long-abandoned Iñupiat sod homes held up by driftwood. For all my stealth, the instinctual animals below had developed a radar far superior to my own. As soon as I poked my head over the sedges, they startled and stomped around into a semicircle, and kicked up a cloud of dust to guard one another in a defensive wagon-train formation and show their would-be predator a wall of curled horns.

After several photographs, and reduced to a shiver, I didn't want to disturb them any longer. As I stood up to leave, I bowed. Then I crept back to my kayak.

The temperature plummeted and as I reached the mainland, the snowy peaks of the Brooks Range and the Arctic Refuge coastal plain faded from view. In place of ancient wilderness, I entered a landscape disrupted by water-filled tire tracks, rusted pieces of metal, abandoned trailers, and pieces of Styrofoam that flew through the air and covered the tundra like errant bits of hail. I had entered the vast, pond-muskeg of America's Petroleum Reserves and the postapocalyptic industrial nightmare of the greater Prudhoe Bay.

While I kept after it with the paddle, I looked everywhere for brown or white or blond boulder-shaped bear movements. I had learned that color and motion are tools to sight life against the threadbare and washed-out tundra. I stopped to sit and eat an Inuit power bar on a two-foot-high by ten-foot-long Mackenzie River tree, whittled down to its prodigious trunk and stripped bare by decades of ice and sea and wind. A polar bear had recently walked the beach with her cubs and imprinted the muddy sand on either side of the log with four-digit toes that poked into the mud and resembled stubby, cartoon fingerprints. Unlike a grizzly, the polar bear's claws are hidden, so the prints looked blurred by all the fur on its paws. It had made a high step over the log without a change in gait. While I could only jump over the high log and leave deep prints on the far side, the long-legged and high-hipped bear came down light-footed in the mud. Her two short-gaited cubs walked around the log with faint prints barely visible. All three sets of prints disappeared into the sea.

Late that afternoon an unseen yellow-billed loon brayed and wailed. As night fell in the oil wasteland, brants, snow geese, and Canada geese murmured, gabbled, and honked. To my ears, in that oil barrens, they could have foretold the world's end.

The next day an altogether different call of a red-throated loon sounded like a wounded child: *Ohhhhhhh!*, while it beat its wings in double time around an oil camp where orange-overalled men refused to return my waves as if they were prisoners in some Gulag. The loon's bill opened and shut as it cried, *Ohhhhhhh, Ohhhhhh, Ohhhhhhhh!!!*

That night the orange glow of a gibbous moon dimmed under the bright yellow oil flares that burned off methane like verdigris against the darkness. Snow gleamed on the shore. A snowy owl hooted. Multicolored pieces of foam from the Badami construction site blew past in the wind. Petroleum products—Styrofoam cups and plastic oil bottles chucked out by Prudhoe Bay workers—wafted closer to my tent as a storm tide crossly beat the shore. I shivered to stay warm.

When I reached the Endicott Causeway the next day, wiped out from whitecaps that had repeatedly breached my cockpit, a security guard waited for me in his heated truck. He got out, in a swagger, heavy-belted with tools of restraint while the truck's strobe-bar lights winked unnaturally blue and yellow as if my mile-an-hour battle against the wind and current out in the bay were cause for a ticket.

I hadn't seen the sun for days. I sat exhausted in my boat, the nerves in my arms numb and jangled from the effort, and stared two miles out to the artificial gravel-island drill pad. (Over the next two years Endicott Oil would be fined $6.5 million for illegal waste disposal there.)[4]

The uniformed man conspicuously leaned a hand on his gun holster. I yelled up, friendly like, that I had just paddled hundreds of miles from Canada.

He yelled down, cop-like, not to come out of the water. A far cry from the kindness of the real North Men.

I just laughed and got out of my boat. It would be a long hitchhike back to Inuvik, Canada.

4 Since oil had been discovered and they began to drill in the late 1970s, the roads and pipelines and oil industry now cover more than two hundred thousand acres across the North Slope of Alaska. Each year across this once-pristine wetland there are hundreds of toxic spills. Air pollution affects villages hundreds of miles away. And beyond the irreparable damages to the land, water, and wildlife, the twenty-five billion barrels of recoverable oil—more than half of it already piped eight hundred miles south to oil tankers—will cumulatively add (as it is burned) over eleven billion tons of carbon emissions to the atmosphere, which is enough to increase temperatures even more around the globe.

Birds
1997–1999

I spent many evenings captivated by a bird guidebook as I tried to reconcile how the northern species I sighted each day looked so different in the book. Each perusal of the birds rendered gigantic and shiny in bright colors on the glossy pages reminded me that, in the real world (but only at first glance), the birds seemed to fade in muted colors as they vanished in the water or sedges.

On closer inspection in the wild, they were entirely different creatures—dynamically and passionately conjoined with land and sea, amid a kaleidoscopic blur of winged color and scattered terra-cotta light—from the guidebook's static images. I tried yet failed to capture them with my telephoto lens.

I paid homage in my journal. How the cry of the sandhill crane—*Grus canadensis* (I wrote and pronounced its Latin name to memorize it)—gave off a bag-of-bones clatter and then a hollow luff of feathers as it shuddered against the air and flew away. Or how the grey plover, *Pluvialis squatarola*, diverted me from its nest and ran comically down the beach, black chest lit shiny as an over-ironed tuxedo under the low sun. Or how the flocked-up mergansers, *Mergus merganser*, with their mullet hairdos and gabble-squawks furrowed the still evening air like lines of surf on a smooth sea.

PREVIOUS SPREAD: With their *trit, trit, trit* vocalizations and aerial acrobatics, bank swallows nest in colonies as far north as treeline in the southern Brooks Range and winter in South America. Iñupiat ancestral lands. JON WATERMAN

Snow geese in flight and sandhill cranes grounded in Bosque del Apache National Wildlife Refuge on the New Mexico Rio Grande flats migrate as far as the Arctic every year, yet another way the North links to the rest of the world. Piro, Mescalero Apache, and Pueblos ancestral lands. JON WATERMAN

Amid the birds of the Arctic, I could commune with the same creatures that migrated to my backyard birdfeeder in the spring and fall and never feel homesick. Birds connect the world together.

I became happily obsessed with birds.

Once I stopped to watch a jaeger (or *isunnak*, long-winged in Inuktitut) hover soundlessly above me. Like the giant fruit bats of Nepal, the jaeger employed swift and authoritative wingbeats. You couldn't miss its distinctive yellow face and formidable wings—nearly four feet long—that held it in place. Its tail feathers worked as a sail, telltale and curled backward as the bird stalled above, then sped off in the wind while those same feathers danced straight back.

Then there was the Sabine's gull that hung in disguise with its black head amid a colony of the remarkably similar Arctic terns with their black-capped heads. The terns sprang up and down and screeched a high-pitched, shrill *kee-ar!* at my arrival. As the terns gave me the usual what-for—with shrieks and dives toward my head—the predatory gull saw its chance and darted toward two fluffy tern chicks prostrate in their nest. Sudden as the wind, a dozen terns swooped in and strafed the gull and drove it back to the waterline. The gull flew off with black-dipped wing edges like racing stripes, but the terns showed even greater agility as they dipped and changed course and drove the chastened gull out of sight.

The migratory champion of all birds, the Arctic tern can log sixty thousand miles a year. The Inuit call it *mitkutailak* (one who goes at the water many times). As a signature species in times of climate change,[1] two million of these quarter-pound birds are found around the world. With their fall migration to Antarctica's austral summer, no animal on Earth sees as much daylight as the Arctic tern. With twenty-four hours of light at both poles, the photosynthetic explosion of plant and small fish growth draws Arctic terns back and forth across the globe like iron boomerangs flung to their magnets.

1 The Arctic tern spends a third of its life on Antarctic ice. NASA estimates that Antarctica's ice melts at 151 billion metric tons per year, so the bird's protective habitat will continue to shrink as sea level rises. The tern could lose 20 to 50 percent of its habitat due to temperature changes linked to climate change (the International Union for Conservation of Nature has "red-listed" the Arctic terns as a Threatened Species). In the Arctic, the terns are likely to lose their nest sites, too, as the loss of glaciers continues to raise sea levels (the loss of Greenland ice sheets will raise oceans as much as twenty feet). The fate of the tern is further compounded by polar bears driven to land to find new food sources (e.g., tern eggs and chicks). Polar bears, deprived of their seal diet, will routinely eat an entire colony's worth of eggs.

When the two chicks that had narrowly escaped predation by the Sabine's gull finished their fledge and took to the air, they would spend several months in flight to Antarctica, with a stop for caloric resupply in the nutrient-rich open waters of the North Atlantic. But on their way back to their Arctic birthplace next April, they would take advantage of prevalent northerlies and zoom home in only forty days.

I never came near the rattle and bustle of a tern colony without a chastened reminder of my place in the world as a ponderous landlubber. I never left without a sense of wonder.

Arctic Solitaire
Spring and Summer 1998

Trip Plan: *In early April 1998, I flew back to Inuvik to continue my journey from where I had started the previous summer on the Mackenzie River Delta. To complete the Northwest Passage—from the Pacific to the Atlantic tides—over the next two years, I would head east across the roof of Canada. Because summers were so short in the Arctic, I wanted to make miles on top of the sea ice before it melted in late June.*

From Inuvik, I hired a taxi to drive me and my dog, Elias, one hundred miles down the frozen river and across the sea ice to the village of Tuktoyaktuk on the Mackenzie River Delta. We drove past pingo hills on the winter ice route (there is no road between the two villages), with thin ice marked by sawhorses. As the orange-barrel route markers disappeared, the African American cabbie—recently immigrated from Somalia—became agitated and drove faster, which sent us into several sideways skids across the

PREVIOUS SPREAD: East of Tuktoyaktuk on the sea ice, I hitched Elias's sled to mine and rode the wind on my skis. The early thaw of sea ice caused by the climate crisis cut this trip short. Inuit Nunangat and Inuvialuit ancestral lands. JON WATERMAN

sea ice. Elias barked in protest. It was the most high-risk and heavily tipped cab ride of my life.

From "Tuk" (population 981), Elias and I pulled our sleds across a peninsula shortcut for two days. That spring and summer, I planned to cover a thousand miles, mostly by kayak, to the village of Umingmaktok.

Atop the frozen ocean, pulled faster than I could run, the parawing kited above me like a gyrfalcon in the wind. My skis skimmed through blue meltwater atop mottled gray sea ice. I leaned back on my harness, counterbalanced by the sled behind me. With weight on my big toe on the left ski, little toe on the right, I steered a slow arc eastward to avoid another open-water lead. I felt tempted to loose a glory whoop, but miles from help out on the frozen Beaufort Sea I concentrated instead and held my core tightly—with my abdomen centered—to avoid a tumble on the lumpy, hard ice. A broken leg wouldn't kill me, but it would be a great inconvenience.

Elias—part rottweiler, part Aussie shepherd—led the way with her own sled of dried caribou, her bed, and a nickel-plated shotgun for bear protection. We had left the village of Tuktoyaktuk on the Mackenzie River Delta four days earlier.

I had trained Elias to skijor. Although she knew the commands to turn right and left (gee and haw), to stop and go (whoa and hike), I brought her for company and for safety because her keen nose and ear would help alert me to bears. At night, she woke up at regular intervals to sniff the air and barked each time caribou came close, leaving me to wake up to clouds of gas so noxious that tears came to my eyes. After years of processed cereal kibble, the *tuktu* meat gave Elias new energy and sent forth a sulfuric miasma.

When we traveled, I harnessed her to her 30-pound kiddy sled and harnessed myself to a 120-pound sled filled with a month's worth of food,

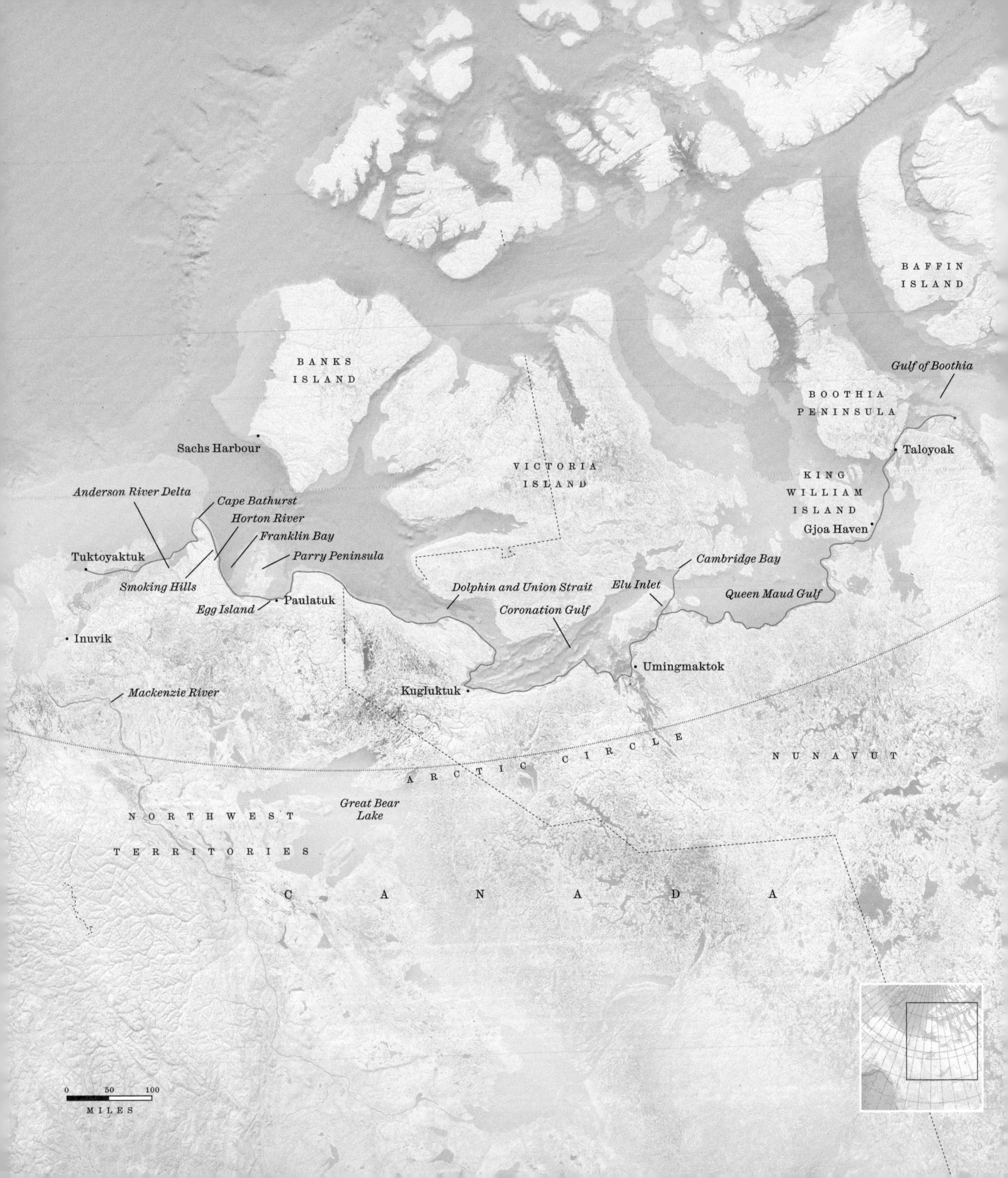

BAFFIN
ISLAND

Gulf of Boothia

BANKS
ISLAND

BOOTHIA
PENINSULA

Sachs Harbour

VICTORIA
ISLAND

Taloyoak

KING
WILLIAM
ISLAND

Anderson River Delta

Cape Bathurst

Horton River

Gjoa Haven

Tuktoyaktuk

Franklin Bay

Cambridge Bay

Parry Peninsula

Smoking Hills

Elu Inlet

Egg Island   • Paulatuk

Dolphin and Union Strait

Queen Maud Gulf

Coronation Gulf

• Inuvik

• Umingmaktok

Mackenzie River

Kugluktuk •

ARCTIC   CIRCLE   NUNAVUT

NORTHWEST

*Great Bear
Lake*

TERRITORIES

C A N A D A

0   50   100

MILES

tent, stove, and fuel that would get us to Paulatuk, the next village. Elias could smell ringed seals long before I could see them, splayed out like shiny brown sacks on the distant ice. Since I had her leashed to me, our pace would pick up as she sprinted toward seal *aglu* holes in the ice. Once, for a brief moment, fifty yards distant, seal and dog faced off while I watched through binoculars. The whiskered seal stared back with lidless spaniel eyes dark as night and a-gleam with the sun; Elias had frozen in place.

The seal gave it up and slithered back down its aglu through the ice with a *kerplunk* of seawater.[1] Elias gave a single bark and looked back at me, head cocked with wonder about what we had just witnessed. (I'd never shared wonder with a dog so regularly as I did with Elias.)

For several days a local Inuit man, Jim Elias, snow-machined out to check on me (his family name may have explained some of his interest in us). After he shut down his machine and pulled out his smokes and lit up, he would ask if I wanted a ride back to Tuk. Each day I politely declined, but I always welcomed his company, however brief, along with the characteristic Inuit thoughtfulness. The last day after he left, we heard the crack of his rifle in the distance as he shot at a seal. Elias gave a quick shiver.

The landfast ice mostly remained frozen tight to the shore, versus the pack ice that floated on the darkened horizon, inviolate and enigmatic off to the north. The water that Elias and I splashed through atop the sea ice had expelled all the salt as it froze last fall. I often scooped the melted sea ice into my water bottle.

Since the wind propelling the parawing had ended, and the sea ice had started a strange, early breakup with temperatures in the high forties, I would soon be forced to walk a roundabout route along the shore. Unlike freshwater ice that begins to freeze at just over 39 degrees Fahrenheit, sea ice doesn't begin to crystallize (and force out the salt) until just below 29 degrees. And while I had walked safely on freshwater pond ice

1 Named for the irregular white-circled patterns in its brownish-black fur, the ringed seal is the smallest and most widespread Arctic seal. They depend upon sea ice for shelter, den holes, and food (often found on the bottom surface of the ice). They dig open and maintain *aglu* breath holes with their clawed flippers and excavate hidden snow caves on top to give birth and hide from foxes and polar bears. In 2012 the ringed seal was first listed as a threatened species in the Bering and Beaufort Seas of Alaska, specifically because Arctic climate change has diminished the sea ice and caused early breakups.

Except for meetings with the remarkable Inuit in villages and hunting camps, the 2,200-mile route I took across the roof of North America proved the most solitary, modern-day journey on the continent.

a few inches thick across the Tuk Peninsula, sea ice remains flexible and unsafe to walk on until it's at least a half-foot thick.

The sea ice beneath me seemed to be solid, but the warm weather had rendered my hoped-for sea-ice snow ski into a water ski—my feet, legs, and sled were soaked. As was Elias. The early spring had also caused coastal streams to flood out over the sea ice.

A few weeks past the yearly maximum extent of sea ice in April, the sea should have been frozen several feet thick. But in spring 1998, open water came much earlier because of a light ice cover the previous fall.[2] And because of the unusual warm conditions and thin ice—due to climate change—I had arrived a month too late.

At seal aglus, or open leads, which we circled rather than jumped over, we looked down into seawater dark as a far-off galaxy with all its inscrutable life-forms. The world had begun to consider the rise of sea levels caused by climate change, but the yearly loss of sea ice (unlike the glaciers that had begun to disappear in Greenland and Antarctica) doesn't raise sea level—just as ice cubes melted in a glass of water wouldn't cause the glass to overflow.

Still, the loss of sea ice has a global reach. Normally, concentrated salt water forced out of the ice sinks and creates currents throughout the Arctic and forces water to move from the poles toward the equator. These currents add to the ocean conveyor belt that influences climate worldwide.

But as the feedback loop amplifies and more seawater is exposed and heated—with the reflective sea ice in decline—habitat for foxes, seals, and polar bears also declines. In addition, as climate change melts landfast ice from the shore, innumerable villages and camps across the North lose their protective storm barrier and are hit by wind-driven waves that cause coastal erosion.

These changes weren't readily apparent as I bungeed the parawing back onto my sled and ski-slogged slowly toward the shore to dry my

2 By September 1998, the minimum extent of sea ice reached a new low record and covered 25 percent less area than previous minimums since 1953. This made it a good year to paddle on relatively ice-free water across the Northwest Passage.

On the sea ice outside the village of Tuktoyaktuk in April 1998, with Elias, about to deploy the parawing to ease the sled hauling. JON WATERMAN

feet and patch my blisters. After only a few days on the sea ice, we had to settle for travel on the land through rotten snow that wouldn't support my weight. So, we dragged our sleds across gravel beaches and in and out of wet snowbanks.

Ptarmigan waggled their clownish red eyebrows and bustled away like white wind-up toys with their curiously feathered, snow-shoed feet, then flushed into the air with their machine-gun-rattle cry: *ah-AAH-ah-AAAAH!*

Out on the sea ice, small bands of caribou wavered up and down in mirages as the land's warmth mixed with the cold sea-ice air.

As the sea ice melts, microorganisms—such as algae, worms, and crustaceans—that cling to the bottom of the ice are set adrift. These microorganisms are the lowest part of the food chain and support all matter of life throughout the North. During every spring sea-ice thaw, the microorganisms draw millions of small, hungry sea creatures—shrimp, sea jellies, krill, and tiny fish—up through the Bering Strait and into the Beaufort Sea. The small sea creatures, in turn, feed hungry ringed and bearded seals, belugas, and bowhead whales. Melt away the sea ice—the first domino in the food chain—and it will then knock over the microorganisms. Then the small sea creatures. Then the whales and seals. Then the polar bears.

After two weeks at a sluggish ten miles a day on a roundabout shore route, with barely 150 miles of progress, I had to quit. In the flooded conditions I couldn't keep my feet dry, which softened the skin and opened raw, bloody sores on my heels. Elias's paws were also weirdly swollen even though she, too, wore protective boots. While I could have avoided the famous grizzly bear population of Franklin Bay with skis out on the sea ice, on land it would've been a foolhardy venture. I didn't want to shoot a bear in defense any more than I wanted to be an offended bear's dinner.

My luck held at the Anderson River Delta. I met a helpful Inuit family— also skunked by the early spring thaw in their efforts to find a polar bear for their "sport"-hunter Cuban American client—and shared a seat out in their chartered Cessna 185. I gave my leftover food to the Inuit, while we crammed our sleds in alongside the rear seat.

Elias and I flew home the next day on Air Canada. Although she flew underneath with the baggage, up in coach I could still smell her caribou gas.

———————

On the unseasonably hot and windy day of June 22, a bush pilot flew me and my Klepper kayak back onto a gravel bar alongside the Anderson River. No sea ice in sight. And for the next ten days—in complete solitude, beyond sight or sound of boats and planes—I battled a headwind that opened the skin on my palms and strained my lower back.

At the tip of Cape Bathurst—a long sand spit that poked out into the sea—the wind finally shifted. Seen through a fog, the sky flushed into an otherworldly periwinkle canvas. While I waited two days for the surf to drop, a blonde wolf trotted into my campsite with tail held high, innocence in her eyes as she loped around me like a skittish dog, just out of reach. Although she didn't stay, through my binoculars I had watched her run two miles out onto the long peninsula to check me out. Like me, the wolf was on a quest, curious, maybe even lonely.

We weren't the least bit put off by each other. I wanted to believe the wolf had sensed something new about me: that I, too, was a creature of the wild. Not a normal human. For days afterward I wondered if the wolf had registered that I had pushed my shotgun to the side as she trotted around me.

When the sea calmed, I put up my sail and clocked many more miles than I would have earned by paddle. In Franklin Bay two days farther along, I wore a bandana over my face and sailed through thick sulfur dioxide clouds that poured out of a two-mile stretch of sea cliffs called the Smoking Hills, where underground shale (exposed to open air as

In light downwind conditions, a jib deployed opposite the mainsail ("wing and wing"
technique) allowed me to sail speedily. I drank hot tea from a thermos and stayed warm
in a dry suit. JON WATERMAN

the permafrost thaws) had burned for untold millennia. Where else could one witness such a spectacle—with the rare mineral jarosite, also found on Mars—but the Arctic? If the Smoking Hills and the abundant bears of Franklin Bay were in the Lower 48, the landscape would've been designated a national park decades ago, but in remote Canada, it remains a distant and unvisited wilderness asylum.

That afternoon I counted nineteen grizzlies. Their sanctuary—alongside the lush Horton River—was vibrant with the chirp of ground squirrels, jumpy caribou, and jade-colored fields of peavine.

Alongside the browned delta I sailed closer to shore to investigate a big bear that patrolled the tundra shelf behind the beach. I greeted it with the usual "Hello, Mr. Bear," which caused him to walk slowly, muscularly, down to the waterline and stand up eight feet tall on his rear feet to get a whiff of me. The fur on his hind end practically touched the ground behind his short legs. As I continued east, he dropped down on all fours and kept pace.

I let the sail luff and stalled the kayak to face him. He stopped again and stood up, ankle-deep in the sea. As water ran off his front paws and he swayed like a hirsute sumo wrestler under his own ungainly encumbrance, he continued to appraise me.

I tried to project my respect across the yards that separated us. Blond highlights in his tawny-brown coat gleamed in the sunlight. He held his fuzzy ears up attentively above his head, with his tiny eyes sunken in a wide, dish-shaped face. My heart beat steadily as the sea pulsed against the shore alongside the bear. For a few brief moments—as the rest of the continent carried on its paved-over bustle and clamor—on the remote roof of Canada surrounded by primeval wilderness and fiery cliffs that belched soot clouds skyward, time slowed and the predator-prey relationship seemed to have been forgotten.

But as the grizzly slowly walked into the sea and began to swim out to me, I had to let go of the fantasy. I had no interest in being reduced into grizzly bear scat—as an Inuit hunter at Anderson River had warned about my likely fate if I tried to walk through the Franklin Bay grizzly enclave. The spell had been broken. I tapped the rudder and pulled in the sheet to fill it with wind. As I gained momentum away, I zoomed back on the bear with the video camera, his head bobbed up and down in the water like a fur-clad jack-in-the-box. (I still wonder what would've happened if I'd waited for him.)

Less than a half million years ago, I knew, the progenitors of this grizzly had transformed into polar bears. Cut off by Ice Age glaciers from the berries, sedges, and tundra-mammal diets, the ancient grizzlies were forced to survive on seals alone. Then they had to learn how to really swim.

Through more than a dozen millennia and innumerable generations, the emergence of the polar bear became one of the most rapid evolutionary adaptations in the animal kingdom. Confined to an icy sea, their skulls began to narrow, their fur whitened and hollowed out for buoyancy, and they grew webs between their toes that made their paws function like flippers. If a polar bear had chased me that day, I would've lost the race.

Despite the power of the encounter, my submersion into Arctic solitaire seemed a greater challenge than curious bears or the thick swarms of mosquitoes. At times, my body actually ached with loneliness as if I were hungry. I took small comfort in the idea that if I were alone and without purpose at home it would have been much worse. I wanted to believe that the solitude would give me new insight into the Arctic and its wildlife.

Distant driftwood on the shoreline morphed into a party of picnickers; icebergs morphed into sailboats. That night I thought I heard a bearded seal talk to me as I tried to sleep.

Along the way I distracted myself with the small but pleasurable routines of mastery. I washed my sails in fresh water whenever I could and coiled the lines neatly and kept the gear lashed in the sail bags whenever the wind dropped. I hunted for wind most every day and even when it eased, I spread the jib sheet wide with my kayak paddle opposite the mainsheet to sail wing and wing in gentle downwind zephyrs. Sometimes I sailed far offshore and perched atop the kayak so I could lean and counterbalance the kayak as it heeled to starboard, free of bugs. Gams of beluga rose slowly in white wavelets that broke the clean surface of the water while their breath hung briefly in the air like the steam of locomotives.

Most days, I leaned back against the heel of my loneliness, and engaged my curiosity and sense of wonder. Then I felt fearless.

When I came ashore and dove into my tent for bug relief, I would squash all the mosquitoes that came in after me. When I went out to cook or pee, I tried to pay no attention to the whine of bugs and to sing; a faux Buddhist amid insatiable mosquitoes that thrust proboscises at me like daggers. More than once I trapped them with a squeeze of my flesh so that they could not withdraw until I released the fold of skin and they pulled out, unable to fly, bloated with blood. *Let that be a lesson!* I thought, as I escorted them back out of the tent.

I filled the map and then my journal with tiny, penciled notes about regal birds and jumpy caribou and the brine smells of the littoral past the sea cliffs burned red and yellow. Or how the peach-lit midnight skies were free of the obese white contrail lines of commercial airliners that avoided this part of the Arctic. I constantly studied the map for whatever new wonders I might encounter beyond the visible horizon, hidden betwixt the contour lines.

The map showed how I could portage across a chain of lakes on the bottom of the Parry Peninsula, which would save me a ten-day paddle in surf around the top. But by the third day of a slog amid the blood-thirsty and incessant drone of bug wings, through an endless kayak drag in lakes gone to bogs, hungry and out of food, my inspired route had turned into a Type 2 sufferfest. Forced to carry the kayak on my back, my spine jangled with an electric ache that made it hurt to stand up straight. Although there were many flightless Canada geese that tottered about in the sedges, even with my gun slung over my back for quick access, I didn't have the heart to shoot dinner.

East of the Parry Peninsula and back to the ocean, I got trapped yet again by the wind on a surf-hammered beach. On my twentieth day alone, I finally met people again. They invited me into their canvas tent on Egg Island. I could scarcely speak after so many days in seclusion but felt in ideal company because my Inuit hosts were equally reticent. Despite the fact that I'd emerged alone from a bug-infested wilderness, across the swamp and over a range of final hills with a disc-crippler, hundred-pound kayak, it would've been considered rude and contrary to Inuit etiquette to ask personal questions about my trip, so they only asked if I had seen any beluga. For once I felt glad to be quiet among other people.

I was now dizzy with hunger, and the elder, Sam, noticed how I ogled a pile of cooked ribs on his plywood floor. When Sam asked if I liked moose, I wondered if he could hear the gurgle of my stomach. Bereft of conversational civility from all the time alone, I replied—a bit too forthrightly—that I ate fish, not meat, but would happily make an exception this time since I was so hungry. Yet again I had been left alone with an elder as if the camp acknowledged that it would be the elder's job to feed and entertain a Qallunaat.

With eyes narrowed, Sam stared at me and tried to figure me out. He had never had a vegetarian emerge alone from the wilderness and come into their whale camp. To Sam my arrival seemed even stranger than that of the moose.

Sam said that he hadn't seen a moose for many years, and although they thought at first that it might have been a caribou, since it didn't run away, he shot it. And since it didn't taste as good as their preferred tuktu, the women in the camp cut it up with their ulus and boiled it for two hours as if it were bear meat. No one in the camp complained because it had been weeks since anyone had seen a beluga or meat on hooves.

Sam got off his chair, bowleggedly rocked across the tent on hips that looked as sore as my back felt, grabbed a plastic plate and filled it with the ribs from the floor, and just as abruptly dropped it in my lap. He lurched back to his camp chair to watch as I picked up the biggest rib and dipped it in a bowl of suspiciously yellowed margarine (that Sam pushed toward me) and bit into the corrugated blackened meat. I had to jerk my head sideways to pull it off the bone.

As I gnawed the leathery meat, Sam held his stomach and broke into laughter. For several minutes he guffawed as I ate; and my jaw ached all the more as tears rolled down Sam's face. I finished the rib, dropped it on the plate, wiped my hands on my pants, and ran my sleeve across my face. Sam offered me a fork, but I asked if he would be offended if I continued without a fork—which brought another round of belly laughter.

With a big gap-toothed smile, Sam said he enjoyed it when vegetarians came to eat moose meat in his whale camp. It was my turn to laugh as I reached for another rib. The meat tasted slightly better than Elias's kibble, smelled of liver, and resisted my teeth like rawhide, but I knew the fat and protein would keep me warm and, I suspected, leave me awake most of that night.

"I don't know why the moose are coming north now," said Sam as he waved his arm toward the shore, "we thought they stayed in the trees."[3]

3   A peer-reviewed 2016 article in the journal *PLOS ONE* cited how moose throughout the North American Arctic had begun to move north of the tree line as the increased temperatures of climate change caused a northward expansion of shrubs (a.k.a. the Greening of the Arctic): "We estimate that riparian shrubs were approximately 1.1 m tall c. 1860, greatly reducing the available forage above the snowpack, compared to 2 m tall in 2009. We believe that increases in riparian shrub habitat after 1860 allowed moose to colonize tundra regions of Alaska hundreds of kilometers north and west of previous distribution limits. The northern shift in the distribution of moose, like that of snowshoe hares, has been in response to the spread of their shrub habitat in the Arctic."

As I paddled away from Paulatuk alone, it felt like every other angst-ridden village departure into solitude. Add to this my overburdened kayak, which plowed low in the water under nearly a hundred pounds of food. Since I couldn't fit a month's worth of food inside the kayak, the aft deck bag was filled with cheese, peanut butter, fruit, and vegetables.

On the bow deck I had lashed down the sail bag with two sails, the three-sectioned mast, outrigger poles, a spare paddle, and a leeboard.

To test tippiness, I braced the paddle on either side to stay upright and rocked my hips back and forth to find the capsize point. Unstable as hell.

Only ten miles from town, a grizzly walked the shore as if he owned it. His huge shoulder hump flexed back and forth with each step like a giant piston. This outsize muscle—unique to grizzlies, attached to their backbone and front legs—allows them to dig out ground squirrels and roots, or execute prey with a single paw swat as if they were bugs. He cast a sidelong glance at me to acknowledge my presence, although he had sensed me long before I had caught up to him. He appeared to be engaged with more important business as he followed his nose toward some scent.

"Hello, Mr. Bear!" I shouted.

He gave me another sidelong look—as if to say, "don't bother me"—but didn't stop. I felt that I had secured an agreement that I would not be his dinner. It was good to get this established, because he could have followed me to my campsite that evening.

I knew that he used his formidable nose to figure me out: the freeze-dried lasagna on my breath from the previous night's dinner, the food in my kayak, the bug dope I liberally applied that morning.

At a steady pace in my kayak, pushed by a current, I passed him. He walked with the distinctive grizzly rear-leg swing that brought the hind paws all the way up to where his front paws had hit the ground. I wished him a good day.

A few miles farther east, I caught up to another narrow-headed, snout-tapered sow bear (obviously followed by the male bear behind us). I confirmed her sex as she urinated with a backward stream. She walked in loose beach sand just as a bear walked in deep snow; she spread her steps out for efficiency and spaced her feet more widely—without the

In warm spring conditions, an Inuit woman from the village of Cambridge Bay at Elu Inlet in the newly made Province of Nunavut (Our Lands) catches Dolly Varden that will be dried on fish racks and eaten throughout the year. Inuit Nunangat ancestral lands. JON WATERMAN

rear-leg swing. As she stepped off the beach and onto a well-traveled route through fireweed and lupine that waved in the breeze, she stepped into old bear prints—to avoid the spongy moss—and followed the evenly placed depressions that were stamped into the moss and would last for decades.

Historically in the Arctic, only the pinks, one of the five Pacific salmon species that attract bears, have been found in limited numbers in the Mackenzie Delta area.[4] While the sea-run, rosy speckled Arctic char and Dolly Varden spawn in northern streams, they don't draw bears like the huge salmon runs in Alaska.

From Denali to the Arctic Ocean, barren ground grizzlies are attuned to musk ox and the movable feasts of caribou migrations. Over the last month, the uniquely white-furred caribou that I had seen along the coast were from the Dolphin-Union Herd.[5] Locals referred to them as island tuktu. Like the northern villagers who mimic predation techniques of the grizzly—which they call *atiqpuq*—the bears will often jump into rivers or lakes and kill caribou in the water (which explained why the grizzly swam after me at the Horton River Delta). Otherwise, on the tundra, bears prey on the slower, sick caribou or newborn calves. The omnivorous barren ground grizzly's mission is to sniff out, hunt down, and consume every available scrap of food.[6]

As the smartest wild animal on the continent—arguably even self-aware—the bears along the Beaufort Sea knew it was safe to be out on the shoreline, away from villages. But if I had an engine on my boat, they would've run off onto the tundra with the knowledge that snow-machine or motorboat engines carried men with rifles. (It's rare to encounter villagers far from their homes—gasoline is expensive in the Arctic and the dogsled is a conveyance of the past.)

Over the weeks that followed, my concern as a trespasser in bear territory demanded vigilance. I set my alarm for regular intervals, woke up and stood outside the tent to scan the horizon (the clear-vinyl windows

4   As the climate crisis has warmed waters and melted sea ice, the chum (dog) salmon of the Mackenzie Delta have recently been joined in the Western Arctic by pink (humpbacks), sockeye (red), and, on rare occasions, the Chinook (king) salmon; there are no records of silver (coho) salmon in the Arctic. Pink salmon are now caught around places such as Paulatuk as harvests in southern Alaska have declined. Still, the spawns are not yet significant enough to attract bears.

5   Recently listed as an endangered species, the Dolphin-Union Herd experienced a catastrophic decline that started in 1997, when the herd that had numbered 34,000 dropped to 3,800 in 2020 due to a combination of climate-crisis-related factors. Unlike other caribou across the continent, this herd relies on sea ice for its summer migration to Victoria Island. As the sea ice has declined, they are cut off from the places they breed on the island. For their migration south to the mainland winter grounds, the uptick in ship traffic along with thin ice—which is broken up by ships and causes many animals to drown—prevents their return. Caribou populations have now declined throughout the Arctic due to the climate crisis, habitat loss, and other issues.

I had sewn into my tent would only have shown a grizzly that was dangerously close). I always slept with a round chambered in the shotgun. Odds were that I would have more close encounters.

For three days a cold front with icy winds pinned me down in a sandstone cove. The next village of Kugluktuk (formerly Coppermine) still lay 355 miles and three weeks away, but storm days would slow me down. I began to ration my food.

Rather than stay in camp all day, forlorn about poor progress, I took long walks to acquaint myself with the land. On these surfy days, amid gusts and sideways rain, the mosquitoes hugged the ground and left me alone. Along the shore, I found engine parts and whiskey bottles and plastic—but in comparison to other oceans, Arctic beaches hold limited flotsam or seashells (most bivalve or shelled creatures can't survive in such cold water).

Away from the beach I discovered a line of rocks a half mile long stacked into several-feet-high, human-shaped cairns, called *inuksuit* (to act in the capacity of a human). A row of ten inuksuit were used to herd near-sighted caribou to the seashore cliffs where the animals could be driven off and butchered by hunters who waited below with ulus and spears.

To go back in time across the Arctic landscape, I wandered far from the sea, like a beachcomber set free. On the rainiest days, I walked in my dry suit. With the shotgun slung across my back, I could defend myself against grizzlies, even though they could ransack the camp in my absence. As I wandered, I pondered bird nests, the cloven tracks of caribou or musk ox, caribou pellets (that told me of recent or long-past animal movement), musk ox fur that waved like prayer flags on the dwarf birch, innumerable animal bones, weathered pieces of *kamotiks* (sleds), partially carved pieces of driftwood, and old footprints of Inuit hunters cast in mud. In the weeks that followed, as stormy seas frequently trapped me on the coast, I walked the land and tried to read it.

6   From 1994–1997 Canadian wildlife researchers gathered and analyzed 169 barren ground grizzly scats to determine bear diets in the central Arctic region that I paddled through. Unlike other barren ground grizzly populations that subsisted largely on fish or plants, they concluded that the bears "lead a predominantly carnivorous lifestyle and are effective predators of caribou" in spring, mid-summer, and fall. In early summer, the bears foraged primarily on green vegetation, and then berries in late summer. "Declines in the caribou population of our study area or long-term absences of caribou may threaten the local grizzly bear population." And, in my opinion, as the Dolphin-Union Herd diminished in the new millennium, it would explain why grizzlies have begun to cross the sea ice or swim out to the islands to seek other food sources—such as seals—and begun to mate with polar bears.

I found many more different types of cairns, such as *inuksuapik*, used to mark campsites, or two separate piles of stones joined together with a pointer rock on top to act as route markers. Some were built with small rock-hole windows on top; when I squared myself up to the window, it allowed me to see an *inunnguaq* more than a mile distant on a ridge. Thick lichen on the north sides of the rocks showed that many of the cairns were ancient. Across the Arctic other cairns have been dated to more than two thousand years old. Only extreme wind or frost action would take down cairns; Inuit would never disassemble sacred inuksuit.

All across the top of the continent, before the European incursion, the people of the North moved and placed stones as lithic memorandums across the landscape. Many were the equivalent of highway signs. Some served as message centers to show where, I imagined, Arctic char spawned, or seals rested, or small game denned. Or where dangerous currents ripped against a headland. Or where friends and family died. Mostly the stacked rocks were unreadable yet artfully constructed mysteries.

I let my imagination guide me with two marker cairns that likely showed a route from the sea to an inland camp. Or another cairn that pointed to a campsite littered with sewing-machine parts and an old gas can, recent enough to show the peoples' acquisition of outside tools and engines yet weathered enough to show how the people—who still spoke only Inuktitut—lived as regal nomads until the mid-twentieth century.

Some cairns appeared to have been constructed as deliberate power loci. Such as a huge pile of rocks (mostly fallen over) where I sat on a day broody with low-ashen clouds that flew like slow-motion birds. I closed my eyes in the apex of this triangle of ancient glacial eskers that spread to the south and resonated with possibilities of wildlife, travel routes, and other secret codices. There, near rocks placed by ancient hands, I felt strangely accompanied, tied to the thoughts and message stones of nomads who passed through over the centuries.

To understand the Arctic, it proved essential to know the people of the Arctic. I had already heard them express their aversion to maps. Yet many villagers took to GPS devices in the same way they eagerly traded rifles for harpoons, outboards for kayaks, and snow-machines for dogsleds. Inuit (or Iñupiat) adapted to these devices as tools that eased their subsistence lives on the land and allowed them to spend more time at home with their families. One can only wonder whether these modern tools—versus their ancient snow goggles or harpoons or igloo inventions—had weakened their bonds with the land and sea and animals. With nature. (But this is the same question that they might've asked me about my handheld GPS.)

Anthropologists believe that the Inuit navigational dexterity is on a par with that of the South Pacific Islanders who crossed the ocean or the Australian Aboriginals who trekked through the trailless desert. But unlike many other Native cultures who rely on Polaris for direction at night, the North Star is not terribly useful to the people of the North since it's directly overhead and isn't seen for much of the starless summer. The same holds with the central Arctic's insignificant tides that become significant rippers in Atlantic-influenced Arctic waters near Hudson Bay, Canada.

Along with their inuksuit, the people still rely on oral traditions and stories for orientation on the land. They employ revered objects that became rooted in their memories, such as a stretch of coastline they call *Ulunguaq* (it looks like an ulu). Or truly humorous references to towers that resemble body parts, such as *Usuarjuk* (small penis). With resemblance to other cultures that have been forced into Christianity and introduced to foreign lifestyles, the establishment of centralized villages ended their nomadism. The loss of their place-oriented Inuktitut language in favor of English—due to the wrongheaded insistence of teachers and ministers—robbed some of their knowledge of the land and the animals. And most of them accepted warm government houses, flown-in food, and welfare checks for difficult lives of survival out on the land.

Inuit hunters rarely get lost. They rely on a constant scrutiny of the horizon for familiar landmarks. Even in fog that blocks a view of landmarks, the wind (although known to constantly shift) provides direction, particularly when gauged by shape-shifted dunes or sastrugi snowdrifts. Animals are also used for directional aids. Birds fly toward or away from the shore, fish swim along it, and seals move toward or away from the ice pack.

Since I passed as a relative stranger through these land- and seascapes, I relied on my maps (my GPS sometimes proved inaccurate in the Arctic). As much as I cherished the maps, they still didn't show the small rise and fall of the tide that exposed uncharted sandbars. Or how currents created new sandbars, or the sea ice that reshaped the coast. Blue lines marked as major rivers on the map often turned out to be dried-up streams. Many islands were unmarked.

As the storms eased, I could put in long days under sail before the wind relented. One day, I traveled forty-two miles. Most days with only my paddle in the wide-hulled Klepper, I felt hard-pressed to cover half that distance. Occasionally the wind blew out of the south for perfect beam-reach conditions, with a warm, licorice-scented breeze that swept over tundra plant blooms and allowed me to take off my wool cap and sit atop the boat and hike my body out to counteract the heel.

I loved how the wind in the sail hummed and trembled throughout the boat while the red, three-inch telltale ribbon on the leech danced straight back to show a perfect trim. When I began to lose speed, the telltale stalled and curled. But correctly trimmed under the vibratory power of the wind, the boat dug in, and I swooped over the crests of swells and zipped my dry suit tight, pulled on my hat, and donned neoprene gloves.

Alongside dolomite cliffs, mustachioed peregrine falcons keened and dove and ripped the air with their wings at reckless speeds. Audibly.

I dropped sail as the wind died and stopped to watch a contradiction of semipalmated sandpipers lost to pleasure as they bathed and preened

Sailing across the wind—more than one hundred miles from the nearest village—sitting atop my kayak as the icepack quelled the surf proved sheer bliss while crossing the Arctic. JON WATERMAN

Weeks out without having seen or talked to people while crossing the vast Nunavut
Province, curious caribou ran into my camp or down to the shore to watch as I paddled by.
JON WATERMAN

in a freshwater pond blinded by a thousand gemlike sparkles, oblivious to my presence, under the most cooperative sun that I could imagine. I stripped to a T-shirt before I resumed paddle work.

To the south, caribou ran in erratic patterns as if in practice to evade a hungry griz. Heat waves shimmered and the tundra seemed to quiver as if it were alive. And the sea hushed into a black quietude of glass as I shattered it again and again with my paddle on the five miles of open water from Coronation Gulf into Dolphin and Union Strait.

For days I became enveloped in a thick, pungent wildfire haze that wafted out of the south as if the world outside had gone up in smoke and the Arctic would be the planet's final salvation. I reveled in it all: the fuzzy black rocks in the water below, the basso profundo gush and gurgle of the cold sea as it beat a rhythm against the sea cliffs, and the Arctic char that flew up out of the water as if to try fins as wings until, subverted, they hit the water again like rocks. I felt attuned to the natural world with an intuition and radar I had not felt before.

One early dawn, a herd of musk ox made their distinctive, guttural growls as they chewed and hoof-raked the tundra right alongside my tent. Surrounded by them, I felt bizarrely fearless, and fell back asleep, soundly guarded.

Caribou ran up to me without trepidation. One night a lone bull came into camp as I boiled water, then trotted away in synchronized splash-hoofbeats through the shallows as if schooled in dressage. Caribou cows, too, repeatedly walked down to the shore—glassy eyed, their backs in repeated shivers with unseen warbles beneath their fur—to study me as I paddled by. One lonely calf sprinted down to the water along a vast, empty shoreline in hopes that I might be its mother, and when I said "Sorry," the newborn stood stock-still and stared with innocent wonder at the sound of a human voice and the paddle that I worked alongside my head.

Bearded seals swirled the water as they swam just below the surface and followed me in my kayak. Now and then, they surfaced to make eye contact. I didn't speak to them if only because I knew that words would undermine these moments of mutual, astonished respect.

Something had changed. *Maybe these wild animals had never encountered people away from the villages?*

One stormy day I chose to battle the waves. I came ashore late at night through the surf to look for a tent site. Away from my kayak, without the gun, momentarily engrossed with a half-eaten jar of peanut butter in my hand, I realized—through neither sight nor sound—that I was not alone on the beach. Amid the vast space and profound solitude that the Arctic confers, when another animal suddenly comes into your orbit, it causes an unmistakable break in your thought patterns. (What else could explain the development of a sixth sense that I had never felt so keenly before?)

I looked back. A barren ground grizzly had just descended from a bluff on the other side of my boat. For the first time in my life, I ran toward a bear without a second thought, fully cognizant of my survival options. If the bear reached my kayak first, I would lose everything.

The bear didn't stop its steady pace, but I reached the kayak first and lobbed the peanut butter jar in its direction. That stopped the bear long enough for me to push the boat into the surf and jump aboard, soaked by a wave. I back-paddled hard. The oddly black-coated grizzly (wet like me, its fur drooped and pigtailed with rain) had its snout in the jar.

Shaken and cold, I continued for another half dozen miles to get some distance. I felt aggrieved to have lost the precious peanut butter, a huge source of needed calories. (I had to cinch my pants tight to hold them above my now-bony waist.) But when I came back to shore two hours later, finally able to relax, the realization hit: I had not smelled, seen, or heard the bear. The predacious mind—a sixth sense, newly honed, or even ancestrally inherited—simply informed me that I had not been alone on the beach.

On the outskirts of Kugluktuk, after twenty-five days alone, I tried to psych myself up to converse and interact with people again. As I saw the first boats in the distance, I practiced a few words aloud.

As much as I wanted to learn from the Inuit, I didn't feel ready to be with people again. It then came to me as an epiphany that, as I lost the angst and nervousness after a week out from Paulatuk, I had found calmness along with a peculiarly aware and instinctive way to be alone in the wilderness.

This state of grace would recur repeatedly, but slowly, after I left the villages. I wanted to believe that the wildlife, too, could read me, and that this could explain the many strange yet peaceful animal interactions.

I had also noted that my close encounters with wild animals never happened in the first few days out away from people, when I carried myself with nervousness and anxiety—gun always held close—in every paddle stroke or step across the tundra. That spring and summer, after a week alone and away from Anderson River or the villages of Tuktoyaktuk and Paulatuk, exhausted from anxiety, fear simply transformed into inquisitiveness. I began to feel the world around me with a new loose-limbed and approachable posture.

This transformation—into more thoughtful and unflappable and instinctively keen behavior—was not the sort of power or mastery that I had expected or sought. But through observation and curiosity I had stumbled across a new threshold.

One Who Gave Power
September 1999

Trip Plan: *It took me ten months, spread over three years, to cross the Northwest Passage. I spent two of those months in early winter 1999 living—rather than traveling—in the subzero villages of Pelly Bay and Gjoa Haven with little sun. In May, I dogsledded with an Inuit man 150 miles from Umingmaktok to Cambridge Bay. When the sea ice went out in July, I continued alone for hundreds of miles in my kayak to Gjoa Haven, then on toward Taloyoak. In 1999, Canada renamed this section of the Northwest Territories Nunavut—which means Our Land—henceforth to be administered by the Inuit.*

I had foreseen the 2,200-mile journey as a once-in-a-lifetime test of my endurance and ability to thrive alone, challenged by errant bears, storms, route finding, bronchitis, and thick hordes of mosquitoes. While the extended solo pushed me hard, every time I emerged in an Inuit village or a hunting camp, I came to a new level

PREVIOUS SPREAD: September 1999, Gulf of Boothia, Canada: For a moment the bear pursued, silent and swift in the water, and I knew I couldn't out-paddle the legendary yet ever-vulnerable creature of the Arctic. JON WATERMAN

of appreciation for the remarkable Inuit. Their perspective on the natural world and its wild animals changed my ways: I learned a new appreciation for true wilderness, I routinely talked to wildlife, I became content to be quiet rather than badger strangers with questions, and I gained a new appreciation for the many privileges and freedoms I hold in my life.

I learned that the best way to understand the Arctic is to spend time with the Inuit and among the wildlife of the North. That was part of another goal for my journey: to meet a polar bear.

In late August 1999, I saw my first polar bear outside Taloyoak village. It cantered across the shore toward the surf like a muscular white pony, hit the breakers a quarter mile away, then sounded and disappeared from view as it swam underwater. If he wanted to catch me, no way I could have outdistanced him. Still, I unfurled the sail and shot across squirrelly waves and watched all around me for the approach of the bear's black nose held above the water—just as he would stalk a seal.

As the legendary creature of the Arctic, the ever-vulnerable polar bear holds a sacred place in our imaginations. At risk of extinction due to the loss of sea ice, the polar bear is reputed as a bloodthirsty killer of humans who venture north—but this is largely myth.[1] Most Inuit hold enormous respect for the bear and often cite its curiosity rather than ferocity. The barren ground grizzly—known to routinely fight and chase off larger polar bears—holds the title for champion aggressor of the North.

It would take me another week to reach the Atlantic tides. Snow geese bound for southern cornfields gabbled in the air above. One flock of

1 From 1870 to 2014, the Wildlife Society documented seventy-three attacks on humans by wild polar bears throughout their range (Canada, Greenland, Norway, Russia, and the United States), which resulted in twenty human fatalities and sixty-three human injuries. "Nutritionally stressed adult male polar bears were the most likely to pose threats to human safety. Attacks by adult females were rare, and most were attributed to defense of cubs." In short, reported attacks on humans were extremely rare.

several hundred passed directly overhead in chevron flight as snow-flakes drifted out of the sky like goose down.

That night with the stars visible in the September sky, I contemplated the age-old belief of stars as holes into the afterworld. When they flickered, Inuit ancestors would knock snow out of the holes down to Earth, while Inuit shamans and polar bears could fly back and forth through the stars. Polar bears were the spirit animals of the North, known as "the ones who give power." Dorset carvers featured them with legs flung out like wings.

In Taloyoak I spoke to an elder who raised his voice in concern when I asked about the polar bear. He shouted that if I met the one who gives power that I should never look it in the eyes or talk to it. And that I should keep my gun hidden unless I meant to kill the bear.

East of the village on my long portage across the tannin-browned lakes of the Boothia Peninsula, several hundred Peary caribou funneled down around ancient inuksuit and into the water. They snorted and swam past with bloodshot eyes, then emerged in a clatter of hooves on the shore where they shook off showers of lake water that blackened the gray stones beneath their white-striped legs.[2]

Back in the sea I dodged icebergs pushed by the wind like ships under sail, their unknowable keels pinioned in swift currents below. I could hear the gunshot-like crack and whooshed collapse and splash of bergs that turned and twisted and splintered in the inbound Atlantic tide. Compared to the pastoral tundra and placid Pacific tides to the west, this was an entirely different Arctic: sullen, wintery boned, and sepulchered in twilight.

A rainbow parhelion encircled the sun as if it were an iris. The ice pack filled the northern horizon like a symmetrical white tsunami poised to inundate the land. Tawny limestone sea cliffs were blackened by tides; leaden-colored verglas coated the boulders above.

2   Found mostly in the Canadian Arctic Archipelago, the uniquely white Peary caribou—a smaller subspecies of the barren ground caribou—doesn't make long-range migrations or live off lichen in the winter like other caribou. Warm rain-on-snow events caused the Peary caribou (found only in Canada) to plummet to 5,400 mature animals in 1996, a quarter of its previous number. The climate crisis has also thinned the sea ice that Peary caribou depend on to cross to the islands. Considered threatened (rather than endangered like other caribou populations), their most recent population estimate is 13,200 mature individuals.

I could hear the gunshot-like crack and whooshed collapse and splash of bergs that turned and twisted and splintered in the inbound Atlantic tide ... this was an entirely different Arctic: sullen, wintery boned, and sepulchered in twilight. JON WATERMAN

Even the miniature Peary caribou here wore icy-colored fur. Arctic char were so thick in the rivers that they could easily be speared. Snowy owls patrolled the littoral like silent drones. Alongside an inlet, I found graves of ancient giants (or so the Inuit hunters there explained the strange human-shaped ovals of carefully placed rocks that marked graves ten feet long).

Anthropologists have studied the remains of an earlier people, the Dorset, who had built stone longhouses and lived in the Arctic for several thousand years before the Inuit and disappeared around 1200 CE. Although no skeletal remains have been discovered, their otherworldly stone carvings are works of art. Many Canadian Inuit believe—unlike the Alaskan stories of the little people—that the Dorset were gentle giants, as strong as polar bears. The Dorset, whom the Inuit call the *Tuniit*, supposedly slept with their legs elevated to drain the blood from their feet to make them light enough to outrun caribou and could drag walruses or lift one thousand-pound seals over their heads.

As I paddled east, seals fearlessly popped their cherubic, rounded faces above the water surface as if to entice several fattened polar bears that lounged on distant ice floes. I kept my eyes on these distant bears—somnolent creatures the color of old piano keys—on sparkly ice pans that undulated in the swells in the vast Gulf of Boothia.

My fingers and toes had gone wooden. My head was filled with anxiety about all the bears.

The previous winter when I had lived with families in Gjoa Haven and Pelly Bay, I learned that Canadian Inuit still legally guided polar bear hunts (the species is listed as threatened and protected by the Marine Mammal Protection Act in the United States[3]). In many Canadian Arctic villages—contingent upon the health of the local bear population—Inuit are issued permits to guide and hunt a limited number of polar bears. These hunts cost the wealthy "sport" hunter clients up to $30,000 and the cash supports the guides, their families, and the

3 The International Union of Conservation of Nature lists the polar bear as vulnerable to extinction, predicted in some studies to disappear by century's end. Of the twenty distinct populations around the Arctic, an estimated twenty-five thousand to thirty-six thousand polar bears remain (census counts are unreliable because the bears live in such remote, hard-to-access corners of the Arctic). Since they evolved to rely on now-diminished sea ice as a platform to hunt ringed or bearded seals, several of these polar bear populations are increasingly forced to live on land. Like fish out of water, and unlike omnivorous grizzlies, many of these carnivorous marine mammals now live on the edge.

villages (the people of the North have never eradicated any of the species that they and their ancestors have depended upon).

Iñupiat stories tell of how the all-powerful Raven god created people and then the animals and plants that they could eat. Then Raven created the polar bear to humble people so that they wouldn't destroy everything else Raven had created.

Humility is what I experienced as I rounded a peninsula of land and paddled too close to a bear in repose: a pallid white beauty alongside her blood-red seal kill. Both lay on an iceberg lapped by the waves, forty yards away and six feet above my kayak.

The bear leaped off the berg and splashed into the water after me to defend its seal. I put on a burst of speed with the paddle, and for a moment the bear pursued, silent and swift in the water. Then in recognition of what I had to do, to show that I wasn't prey, I pushed the rudder, braced the paddle, and spun the kayak to face the bear, but without eye contact. I used a careful and respectful peripheral glance to point the camera and fire off several photographs.[4]

4 When I developed the photographs later, one showed the polar bear's mouth held open and her ears held high, which classically signals anticipation. Ears flattened would've signaled aggression, which precedes a charge or attack.

I didn't speak. My gun was hidden.

She circled back to the iceberg. I put my paddle down, gave a long bow, and kept my eyes focused low, to the water. Then I turned the kayak to head back to Taloyoak. Even if my Arctic education was still not complete, I had gotten all that I had come for. The polar bear encounter completed my Northwest Passage.

From a distance I looked back and the one who gives power had vanished.

Refuge
Summer 2006

Trip Plan: *I had pestered National Geographic to support my journeys across remote landscapes in the North for decades. In 2006, they acquiesced. My goal was to celebrate the fiftieth anniversary of the 1956 scientific fieldwork that led to the creation of the Arctic National Wildlife Refuge and to study the climate crisis. I arranged a summer's journey in northern Alaska with one of the graduate students from that long-ago study, Dr. George Schaller—now a renowned field biologist and a revered friend.*

Over two months, with three graduate students in tow, we interviewed scientists and Indigenous people in Fairbanks, the Toolik Field Station, Prudhoe Bay, Kaktovik, and Arctic Village. Along the way, we climbed some mountains and floated the Marsh Fork of the Canning River, in the western Arctic National Wildlife Refuge.

PREVIOUS SPREAD: Permafrost thaw on the Beaufort Sea bluffs, with the highest erosion of any coast in the world, due to warming temperatures and a lack of sea ice. Arctic National Wildlife Refuge, 2006. Iñupiat ancestral lands. JON WATERMAN

*We also visited the headwaters of the Sheenjek River
in the southern refuge. This is where George began
his extraordinary career a half century earlier. As he
returned home, I would embark on a 175-mile traverse
detailed here—on foot, by packraft, then in a kayak—
from the Sheenjek over the Brooks Range to the coast of
the Beaufort Sea, then on to the village of Kaktovik on
Barter Island.*

While the polar bear has become the poster child for the climate crisis, the Porcupine Caribou Herd in Alaska has come to symbolize what could be lost to oil development in the Arctic National Wildlife Refuge. But northeastern Alaska only represents the tip of the iceberg when it comes to future harmful industries—oil fields, open-pit mines, and ship transit—across the wild and undeveloped circumpolar Arctic.[1]

I returned to the refuge to weigh these potential damages seven years after I completed my Northwest Passage journey. Although the glaciated peaks here are some of the highest in the Brooks Range, to their north is a coastal plain bypassed by the last Ice Age.

I planned to traverse thirty miles from south to north over the range. Then we would inflate our packrafts and paddle twenty-five miles down the Kongakut headwaters to meet a third friend with a cache of kayaks, and paddle seventy miles to the Beaufort Sea. Finally, we would thread fifty miles of icebergs west along the coastal plain (that I had last passed in 1997) to the Iñupiat village of Kaktovik.

Half of the residents are in favor of proposed development on the adjacent coastal plain because Prudhoe Bay oil brought royalties, jobs, and plumbed water to their village. But as the subterranean dipstick has lowered over the decades, the petroleum riches of Prudhoe Bay weren't enough for the oil companies.

1   In 2008, the U.S. Geological Survey (USGS) conducted a study and concluded that "the extensive Arctic continental shelves may constitute the geographically largest unexplored prospective area for petroleum remaining on earth." The USGS estimated that up to a quarter of the planet's undiscovered and recoverable petroleum resources lay in the Arctic: 30 percent natural gas, 20 percent liquefied natural gas, and 13 percent oil—all mostly offshore.

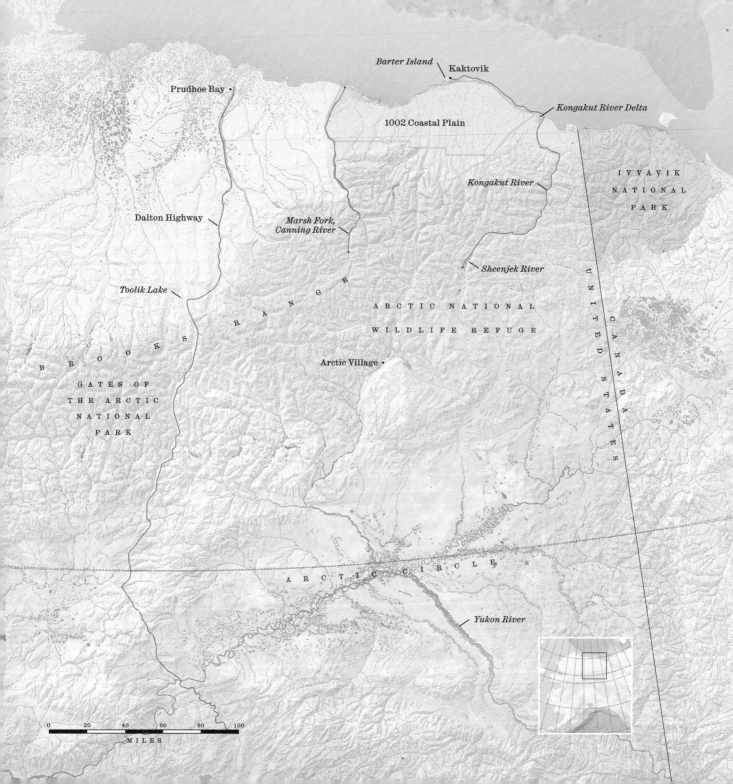

BEAUFORT SEA

Prudhoe Bay •

Barter Island — • Kaktovik

Kongakut River Delta

1002 Coastal Plain

Kongakut River

IVVAVIK NATIONAL PARK

Dalton Highway

Marsh Fork, Canning River

Sheenjek River

Toolik Lake

B R O O K S   R A N G E

A R C T I C   N A T I O N A L

W I L D L I F E   R E F U G E

U N I T E D   S T A T E S

C A N A D A

GATES OF THE ARCTIC NATIONAL PARK

Arctic Village •

A R C T I C   C I R C L E

Yukon River

0 20 40 60 80 100

MILES

Eventually more pipelines tentacled east and west of Prudhoe Bay along with an array of pump stations, gravel roads, refineries, and buildings. All lay on the biologically productive coastal plain that Iñupiat throughout the North Slope knew as a sanctuary for birds and caribou. Since time immemorial Iñupiat have subsisted on these caribou and based their cultural identity on the hunt.

In 1980, Congress called for a study that would either open the refuge's 1.5-million-acre coastal plain (referred to in government documents as "1002") to oil leases or protect it forever—like the rest of the refuge— as wilderness. Well past the new millennium, in what amounted to a forty-year stalemate, Congress continued to float bills for development, countered by bills for wilderness protection. It would appear, at least for now, that neither side has won.[2]

With the decline of sea ice, Beaufort Sea polar bears have often been forced to come ashore and den on the narrow strip of coastal plain that stretches from the refuge and widens out to the west across the top of Alaska. Over the millennia the coastal plain has been a restorative kitchen and migratory rest stop for hundreds of thousands of caribou and lesser snow geese, among many other species. So Big Oil has pitted profit incentives (with the rationale that production will fulfill energy demands and national security needs) against beleaguered wildlife and wilderness—all heated up by the climate crisis. And while the oil lobby has claimed development doesn't harm wildlife or wilderness, one only has to spend a day in Prudhoe Bay to see the destruction the industry has caused locally. Never mind the oil field's contribution to greenhouse gases that continue to worsen the global climate crisis.

In late July, the photographer John Burcham and I climbed from the canyon floor of the Sheenjek River in the southern refuge up onto a tableland of sedges bent double in the afternoon wind. Nearby a large grizzly worked the ground squirrels and peavine as we climbed a ridge

[2] The 2017 Tax Cuts and Jobs Act, enacted under the Trump administration, enabled the sale of oil and gas leases in the refuge. But in anticipation of the expense of development, along with lawsuits from the environmental coalition, none of the major oil companies stepped up. Only the state of Alaska purchased a half dozen leases to drill. Then in 2020, the Biden administration suspended oil leases in the refuge and revived the decades-old battle between the oil industry and environmentalists.

In 2006, with the wildlife biologist George Schaller, we celebrated the fiftieth anniversary of his field study of the Arctic Refuge and investigated climate change at Toolik Lake, Prudhoe Bay, Kaktovik, and Arctic Village, and on two river journeys through the wilderness.

toward the north side of the Brooks Range. We were both, as the poet Robert W. Service wrote about his shotgun-equipped miners, "armed for bear" with pepper spray.

For my first visit to Alaska in 1976, my father had given me Service's book *Spell of the Yukon* (Dad had composed piano music to the poet's work). The rhymed prose-poetry told of wild-eyed prospectors who abandoned their mines and found something more valuable than gold—a peace of mind—among nameless Arctic mountains and "valleys unpeopled and still."

Service hadn't been my only inspiration. I'd spent the previous month with the renowned field biologist George Schaller, my professor friend Gary Kofinas, and three grad students as we toured the North Slope, visited with scientists, and traveled throughout the refuge. Our summer's travels would be both a celebration and a re-creation of George's time in the Arctic a half century earlier.

As a grad student, George had spent the summer in northeastern Alaska and worked on a biological survey that led to the creation of the Arctic National Wildlife Refuge in 1960.[3] In his long career, Schaller saved more wildlife species and habitat than any conservationist. The list includes mountain gorillas, snow leopards, jaguars, Tibetan antelopes, giant pandas, and Marco Polo sheep in Nepal, Brazil, Pakistan, Mongolia, and China.

With our boots on George's trail, John and I continued on our "turf and surf"—from the high tundra to sea level—journey. In a remote drainage north of the Sheenjek River, caribou trails lined the valley like lifelines on a giant's palm. We followed bear tracks along a mossy creek laden with bones, tracks, and scat. Waist-high willows tangled the creek bottom, but caribou and bears had stamped out trails in any direction we cared to walk.

Several weeks earlier, with grad student Forrest McCarthy, I had climbed a peak on the north side of the Brooks Range to recreate a 1910

3   Schaller's 1956 scientific trip was led by the legendary field biologist Olaus Murie and his wife, Mardy (later known as the grandmother of American conservation). Although they had heard the area held diverse habitats and wildlife, what they recorded astonished them: 86 bird and 19 mammal species, as well as snails and numerous insects (today the US Fish and Wildlife Service has documented more than 200 bird and nearly 50 mammal species). "My vials," Schaller wrote later, "contain 23 spider species, to give just one example of the invertebrates." He collected 40 species of lichens and 138 kinds of flowered plants. Olaus shared the secrets of diligent fieldwork that George took on as he followed animal tracks up over the Brooks Range to the north side, while he collected and pulled apart innumerable scat and focused on everything from large mammals to the Arctic poppies that bloomed at his feet. Olaus showed Schaller and the other graduate students how to meld morality with science and to lobby for issues they believed in.

During its migration in the southern Arctic National Wildlife Refuge, the Porcupine
Caribou Herd follows its ancient skein of trails up to breezier high ground for insect relief.
Gwich'in Nàhn ancestral lands. JON WATERMAN

Above the Sheenjek River in 2006, the wildlife biologist George Schaller follows Forrest McCarthy up a peak he first climbed fifty years earlier, during the scientific study that led to the creation of the Arctic Refuge. JON WATERMAN

black-and-white photograph of a more heavily brushed creek valley. Our contemporary view, as we held up the old black-and-white photographs, showed a loss of leftover winter overflow ice, or aufeis, in the river (but this could have been a seasonal variation). More to the point, and what we hoped to prove about the Greening of the Arctic over the last century, willows in the creek drainages below had grown to ten feet tall and flourished throughout a now-warmed plain that in 1910 held only bare tundra.

On the trek with John, our tundra walk out of the upper Sheenjek bore all the familiar aspects of Arctic travel, through wide-open spaces akin to the steppes of Mongolia, under a gunmetal overcast that muted the sun. A golden eagle plied the updrafts, while a caribou bull kicked through unstable talus rocks with the clamor of loose plates in a dishwasher's rinse cycle. That night the sun shot its brilliant rays beneath the clouds and tinted our yellow tent walls mango.

On our third day, with our hoods cinched against an icy rain that blew off the Beaufort Sea, we crossed an unnamed five thousand-foot-high pass into the Kongakut River headwaters. The treeless landscape abounded with orange lichens, shed caribou antlers, and a profusion of brilliant yellow Arctic poppies that opened and closed their hoods and slowly rotated under the movement of a meek, half-hidden sun. Clouds hung on innumerable gray ridges alongside stream valleys that added their flows to the Kongakut River as it twisted and curled past minarets and limestone towers.

The cool weather had forced the mosquitoes to lie low as we found the first burble of the Kongakut. Our clothes sagged with rainwater while we trudged beneath well-traveled Dall sheep trails, etched white into the rust-colored mountainsides above.[4] Our hearts leaped in our chests as ptarmigan clucked explosively out of the brush in a white whirr of wings with downy feathers that floated to the ground in the birds' slipstreams.

4 Although the Dall sheep population (roughly fifteen thousand) is presently stable in the Brooks Range, their numbers have declined dramatically in southern Alaska. The climate crisis has caused shrubs to move up the mountainsides and replace the sedges and other growth the sheep depend on.

After we finished the last packets of oatmeal the next morning, we marched downstream as the threadlike headwaters doubled, then tripled in volume—blown up by innumerable streams. When the river had deepened enough, we inflated our four-pound packrafts, assembled collapsible paddles, zipped into personal flotation devices (PFDs), lashed our packs to the bows, and jumped in. Freed from the stumbly passage through tussocks and boulderfields, we whooped with joy.

The current flushed into a waterslide as we pushed off banks that dripped with shiny gray permafrost. Grayling darted wraithlike through eddies dappled in golden light. A biophony of bird calls—thrushes, warblers, and chickadees—rose above the throaty gurgle of the current. From the western bank we smelled the sulfurous-methane tang of ripened bowels, so for a good half hour, in deference to a caribou killed by a grizzly, we put our heads down and paddled hard.

We darted left and right all day through a labyrinth of river braids. The channels often shallowed out alongside aufeis that steamed under a warm sun and forced us to jump out of the boats and walk our rafts like dogs heeled on their leashes as we splash-walked to deeper water. That evening we reached Drain Creek, the normal put-in for Kongakut River runners, alongside a lumpy gravel bar utilized by small Cessnas equipped with fat tundra tires.

Mike Freeman, our third companion, welcomed us exuberantly. He'd spent a lonely six days stationed alongside the river with our three kayaks and two weeks of food. While time alone in the Arctic can be a meditative and even curative experience that offsets the pressures of the outside world, companionship in the immense wild spaces of the North can be safer, more relaxed, and even fun. Or so we agreed as Mike filled our cups with red wine, and we recounted the last week's experiences.

In the morning, we sped downriver. Birds erupted from the willows. As the river valley widened into a rich green plain below the foothills, two Dall sheep bolted off their graze and up onto the safety of limestone cliff ledges. One of the rams sported a full curl of horns and lay down a hundred feet above to chew his cud, satisfied that he was safe from the interrupters of his dinner.

It took another three days to reach the broad delta. As we approached the sea an icy wind sat down hard on the river and lifted glacial till off the banks that blurred our vision and stirred the vast landscape into a mystical, dusty pandemonium.

When it stilled and warmed that afternoon the coastal plain air passed over the icy sea and bent the light waves into the mirage known as Fata Morgana (the treacherous fairy sister of King Arthur). In the distance, an ocean liner or oil tanker wavered up and down alongside enormous sapphires.

"Just a mirage," I reassured John and Mike, both new to the Arctic, albeit initially dismayed that such a huge ship would be so close to the wilderness.

We threaded our kayaks through six-foot-high pans of aufeis cloaked in brilliant mist and put on our sunglasses near a caribou carcass, dropped in its tracks from old age. Hundreds more caribou trotted back and forth in crunchy steps across the bug-free ice pans and scattered as we approached.

The Porcupine Caribou Herd traveled up to 1,500 miles in their round-trip migration from the Porcupine River area forests to the coastal plain on either side of the international border. Each summer they aggregate into a tight mass of over a hundred thousand animals out on the windy plain to succor their calves in relative freedom from insects, golden eagles, wolves, and bears. Predators still manage to kill up to a quarter of the calves each year.

If an oil field were built on the plain west of the Kongakut River, biologists believe that the herd would be driven into the foothill terrain of their predators. Off the coastal plain, the caribou would suffer greater

I paddle along the landfast sea ice of the Arctic Ocean that clings to shores and prevents
village flooding and surf erosion of shorelines. This ice is disappearing because of climate
change. Iñupiat ancestral lands. JOHN BURCHAM

losses that would eventually decimate the herd.[5] Then the heart of this remote wilderness—as large as South Carolina, but with fewer than a thousand visitors a year—would endure the methane reek and unnatural light of gas flares, the clang of oil rigs, dusty roads, and the glint of pipelines.

If the plain were developed, the United States would gain less than a year's supply of fuel over an expected half century of exploitation, while oil contractors would strike it rich. In turn, the Sacred Place Where Life Begins (as it was named by the Gwich'in of Arctic Village) would go bankrupt.

So, we imagined it as we paddled into the coastal fog. Visibility shut down as we paddled out of the river and into a coastal lagoon. I dipped my hand into the water and licked a finger: still fresh, which meant that the tide was out.

Another mile to the northwest the fog burned off as we reached a sandy island, Icy Reef, washed out of the refuge by the myriad of rivers. We shimmied out of the kayaks and stood up to shiver with our mouths agape: Forty miles south, the Brooks Range rose above the verdant coastal plain. To the north, the mirage ship appeared again, its red hull now clearly visible beneath a black cloud of diesel smoke, surrounded by icebergs in the Beaufort Sea. We could hear the rattle and clang of chains with the *whoomph-whoomph-whoomph* of piston machinery— this was no mirage. Shell had been authorized to drill offshore in 2007, and this exploratory ship had come to map the potential prospects— despite the damage that oil spills, or the considerable noise and harm from the ship's seismic guns, could cause to polar bears, seals, whales, and other marine life.[6]

We camped on Icy Reef, irritated by the noisy ship offshore. I tossed and turned in our big tent that night, piqued that—even with the fate of the refuge still undecided—we couldn't escape the long arm of Big Oil.

5 Of the major caribou herds in North America, only the Porcupine Herd has stabilized and grown. The latest census in 2017 counted 218,000 animals—forty thousand more caribou than its peak population in 1989.

6 These vessels use seismic air guns, which continuously blast loud, low-frequency sound waves through the water column and into the seabed. The operations can go on for weeks on end.

Shell attempted to drill immediately offshore in 2007, but accidents and ship crashes in the poorly charted Arctic waters caused them to shut down operations. The company then moved its offshore ship platforms— replete with crew quarters and helipads—to remote locations in the Chukchi Sea but, after investments of several billion dollars, were unable to find oil. So, they shut down in 2015. In 2020, the company announced that it would begin offshore oil exploration again adjacent to Prudhoe Bay.

In a morning covered with frost, we continued our paddle east through calm lagoons, past Iñupiat grave markers catawampus in the thawed tundra. We stopped at the ruins of an Iñupiat sod-and-driftwood home, abandoned sometime after the Cold War when the people were forced into villages and ended their nomadic lives as traders along the coast.

Farther east we paddled past bluffs that leaked silty permafrost water onto the sandy shores. For miles, the coastline slumped in mud quagmires. In places, the thawed-out land had split open with deep fracture lines, as if the ground had been shaken by an earthquake. Elsewhere the storm surf—unchecked by the recent loss of sea ice—had begun to erode the ocean bluffs (nine years earlier when I'd come through, many of these bluffs had been intact and frozen solid).[7]

On our last night out, the wind came off the ice pack and we took shelter behind stacked logs under a bluff. With a match and dried-out grass, John blew a fire into life with Mackenzie River driftwood, blown west from hundreds of miles away.

As at the end of every good journey, we talked of the joys of our trip: the Dolly Varden we had caught, or the capsizes that didn't faze Mike in his dry suit.

I talked about how George Schaller had inspired me with his ever-present journal, tucked into his breast pocket. Or how he yanked strings out of his pants pocket to measure the girth of cottonwood trees that we believed had spread north and flourished in warmer temperatures.

What I admired most, I said, was that even as a renowned scientist, George always spoke up on behalf of the environment. After our trip, George lectured about wilderness protection for the coastal plain to back-to-back audiences that overfilled the auditorium at the University of Alaska Fairbanks. He believed that anyone who witnessed the destruction of the environment had no choice but to become an advocate for conservation. To take action and speak out.[8]

7 From 1955 through 2015 on the nearby north shore of Kaktovik's Barter Island, the bluffs have retreated up to 433 feet. In a single year between July 2014 and July 2015, the bluffs retreated an average of 4.27 feet, and up to 26.5 feet. Most of the bluffs degraded in summer through a combination of Beaufort Sea storms (the land now often lacks protective sea ice) and permafrost thaw.

8 From 2003-2006 Congress considered oil development in the Arctic Refuge within budget resolutions, or defense authorization and energy bills. After my trip, I spent the fall and winter of 2006-2007 on a public lecture circuit to scores of venues around the country with stamped envelopes addressed to the appropriate members of Congress. I helped my audiences write to their representatives and ask them not to support development. This take-action campaign against the billion-dollar oil lobby—like David against Goliath—proved to be one of many nationwide outreaches that helped defeat a bill that proposed oil development in the refuge.

On that hushed August night alongside the coastal plain, as icebergs pushed against our islet, we crowded closer to the fire. Amid this land- and seascape of continuous summer daylight, the newly arrived night deepened to a dark blue velvet. Ursa Major—the cluster of stars that forms the great bear—glimmered and commanded the heavens with the remarkable power that I had once felt and come to expect from a polar bear.

It seemed only appropriate then to recite my favorite passage from Service's *Spell of the Yukon* about a miner who'd given up the gold:

"It's the beauty that thrills me with wonder,

It's the stillness that fills me with peace."

Shocked Return
August 2021

Fifteen years later, back on the Noatak River in Western Alaska, with my son Alistair; he had never been anywhere like it. I had brought him and a family of friends for a float in the headwaters. What we experienced over the next week shocked me.

My trips in the 1980s were to a colder, less brushy place filled with caribou. This trip we saw just one.

Three decades earlier, the river headwaters were often sluggish and slowed by huge meanders. But in the first week of August 2021, every potential campsite had been sluiced over with silt and mud by the extreme rainfall that had lifted the river out of its banks. I had to keep a steady pull on the oars to stay midstream so the unusually strong current didn't pull us into the banks.

The flooded river pushed our raft and inflatable canoe downstream into a brisk wind as cumulus clouds gathered like ripened fruit above. We had come in August so that frosts would quell the infamous mosquitoes. But since the climate crisis had lengthened the summers and delayed the winters, we needed head nets and bug dope (three decades ago Augusts were not buggy).

PREVIOUS SPREAD: Alistair Waterman contemplates the vast spaces of the Arctic, amid a flooded river that washed out campsites and gravel bars throughout the Noatak headwaters. Dënéndeh, Kuuvuan KaNianiq, Gwich'in Nàhn, and Iñupiat ancestral lands.
JON WATERMAN

Since the pandemic kept most people at home, we would only see one other boater. And while our group—new to the Arctic—had been filled with trepidation, they quickly developed a new awe and appreciation for the unusual landscape that unfolded before them.

Fifteen-year-old Alistair looked on with eyes of wonder. I taught him how to expel breath from his lips to make the sharp pish calls that attract birds, and we called in warblers, redpolls, and juncos. We identified tracks—caribou crescents, fox paws, and the V toes of sandhill cranes—imprinted like art on the riverbanks. I taught him shotgun safety.

On the third day out a grizzly sow and cubs cut off our return to the tent, so we sat back and watched, unseen by the bears, as they gamboled and splashed through the tundra a few hundred yards below us. After they passed, I picked up and pulled apart their scat that showed a meat-free diet of sedges and berries.

The most powerful experiences a parent can give their children—amid a complex information era and an urbanized world disconnected from the outdoors—is to unplug and share the beauties of wilderness filled with animal life. I wanted Alistair's senses to open up with appreciation and let him find a sense of place and purpose amid a time of chaos and climate crisis.[1]

We had also come to escape the record heat and forest fire smoke stateside for what I believed would be a cool interlude in the Arctic. Yet to my surprise the temperatures on the Noatak hit the nineties, so Alistair and I repeatedly cooled off with river swims, which I hadn't done in dozens of trips to the chilly North.

When a storm blew in on August 5, temperatures dropped into the fifties and rain fell again as we floated out of Gates of the Arctic National Park and into the Noatak National Preserve. The legislated wilderness shared between these two parks stretches more than thirteen million acres, which makes it the nation's greatest wilderness area, as well as the largest unaltered river system. But given the cascade of climate anomalies, the region's protected status seems scant consolation.

1 The distinguished biologist E. O. Wilson explored the idea that humans have an innate relationship with nature. In his book *Biophilia*, he posited that lack of exposure to the natural world has caused us to begin its destruction, ecosystem degradation, and species loss (he called this the "machine in the garden," although the climate crisis had scarcely been noticed when he published the book in 1984). When it came to the environment, Wilson wrote, "Our sense of wonder grows exponentially: the greater the knowledge, the deeper the mystery and the more we seek knowledge to create new mystery."

With the raft propped up for sun protection in the Noatak River headwaters as temperatures soared in the ever-hotter Arctic summer, Alistair cooks grayling for dinner after he took a cooling swim in the river. JON WATERMAN

The more I thought about it, the longer the list of changes grew. Our water filter repeatedly plugged up from the dislodged sediment.[2] The wolf den we had discovered in 1983 on a high riverbank of the Noatak amid knee-high dwarf birch and sedges was overgrown with head-high willows. I shared with Alistair that my feet mostly stayed dry on my tundra hikes in the 1980s, but on this trip, we repeatedly soaked our boots on tundra drenched with thawed permafrost. The mountains were also strangely bereft of snow.[3]

Amid all the changes we discussed—permafrost thaw, the lack of caribou, new shrubbery, warm temperatures, and diminished sea ice—Alistair expressed sadness for the state of the world. I replied that a good cure would be to learn more about the extent of the damage and then to take action.

Unable to sleep our final night at our takeout alongside Lake Kavachurak, I listened to a Wilson's warbler sing—a sweet and accelerated aria of notes that descended the scale. I slipped past Alistair, sound asleep, and emerged from our tent into the surreal soft light of a midnight sunset. The lake held the vacant blue of the sky. I climbed onto the headland above as a rainbow arced in a complete bridge over the river. As much as anything, I realized, it was the light that had drawn me back again and again. The winters I had spent in Alaska in low light were often a challenge, but when the light returned in spring and stayed up most of the summer, I felt new energy, like the birds that sang through the night.

North of the Arctic Circle in summer, late at night you could wear this light on your skin like the finest garment; you could bathe in it; take pictures in it. And gyrate through it, as the Arctic poppies do.

I wanted to continue downriver, for weeks on end. And get lit.

But as I stood under the rainbow, I worried about the world that both my young sons would inherit. Will they lose this wilderness to a great thaw?

I knew that I had to come back to document the change—to continue the length of the river and learn what might be lost.

[2] A recent study of the area's smaller rivers and streams found that the permafrost thaw cooled the waters, which biologists say could hurt salmon reproduction. This finding raises long-term concerns for remote communities downstream that depend on salmon for sustenance.

[3] Year-round snow cover in Gates of the Arctic National Park has nearly disappeared. As per one study, thirty-four square miles of white snowfields were seen in 1985, but by 2017 just four square miles remained.

PART TWO

The Final Journey
2022

Descending from Peak 4880 back to camp on the Noatak headwaters amid wildfire smoke that blanketed the Arctic in the summer of 2022 and burned over three million acres. CHRIS KORBULIC

A Short Walk over the Brooks Range
July 12–15, 2022

I lick my lips and get a quick taste of bear spray that wakes me up with a start and reminds me where I am: 3,900 feet on the crest of the Brooks Range about to cross over into the seldom-visited source of the Noatak River. I slept like an exhausted hibernator, despite the gusts of wind that flopped the tent bug liner in our faces most of last night. Chris and I lie atop a lumpy, tilted bog that drains to the lake.

I groan when I move my hand and tweak my thumb, which shoots an electric jolt up my arm. It'll be hard to clutch a paddle for weeks on end. I've worn out the cartilage that cushions the bones in my right thumb.

Chris fills a pot with water as I light and pump the stove. I stumble out of the tent to pee and am immediately in pain—it feels like my urine is a stream of pepper spray that will burn a hole through the live mat of tundra. On the bright side, the sun is out and it didn't snow last night, so I am relieved in another way: we won't have to break trail.

Ice has glassed over the lake edges. Clouds scud by. It's a day worthy of my down jacket, which I didn't bring for this short walk over the Brooks Range. A misguided concession to a lighter pack, so I'll have to jog in place until the day warms up.

Our 2022 route took us five hundred miles over the Brooks Range and through three National Park Service units and villages on the Noatak River and up the Chukchi Sea coast.

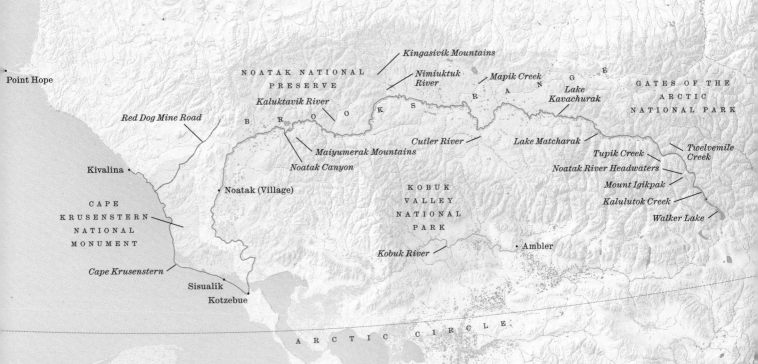

CHUKCHI
SEA

Point Hope

Red Dog Mine Road

NOATAK NATIONAL
PRESERVE

Kingasivik Mountains

*Nimiuktuk
River*

Mapik Creek

Kaluktavik River

B   R   O   O   K   S   R   A   N   G   E

*Lake
Kavachurak*

GATES OF THE
ARCTIC
NATIONAL PARK

Kivalina •

Maiyumerak Mountains

Cutler River

Lake Matcharak

Tupik Creek

*Twelvemile
Creek*

Noatak Canyon

Noatak River Headwaters

CAPE
KRUSENSTERN
NATIONAL
MONUMENT

• Noatak (Village)

KOBUK
VALLEY
NATIONAL
PARK

Mount Igikpak

Kalulutok Creek

Cape Krusenstern

Walker Lake

Sisualik •

Kobuk River

• Ambler

Kotzebue

ARCTIC   CIRCLE

*KOTZEBUE
SOUND*

BERING
LAND BRIDGE
NATIONAL
PRESERVE

0   10   20   30   40   50

MILES

I look up at the mountainous skyline and take a deep breath of appreciation. We are in the northernmost and least visited of all national parks. The annual visitation figures released by Gates of the Arctic National Park and Preserve are grossly inflated. Most of last year's seven-thousand-plus visitors never entered the park; they were counted at the distant visitor center on the dirt highway in Coldfoot or outside the park at the ranger station in Bettles. No way you can find a real wilderness experience amid the crowds in most national parks.

At a lodge in Bettles, Chris and I overheard two couples at an adjacent breakfast table talk about their quest to visit all sixty-three national parks. Their plan that day for Gates was to fly over a corner of the park in a small plane, take a few pictures, and get the hell out. It's no secret that on the ground you can easily get lost in the massive wilderness, filled with bears and mosquitoes.

As the second-largest national park—it's almost as big as Belgium—Gates contains more than eight million acres of uninhabited tundra, scraggly boreal forests, thousands of streams, and who-knows-how-many peaks. Gates is also adjacent to four more protected landscapes managed by the National Park Service that total eleven million acres. Chris and I will pass through two more—Noatak National Preserve and Cape Krusenstern National Monument—on our way downriver and along the Chukchi Sea, a five-hundred-mile-plus, month-long, summer-vacation tour to the Iñupiat village of Kivalina.

Unlike most parks, Gates has no roads,[1] campsites, trails, buildings, or cell phone reception. Although it would be improbable for us to encounter park service folks in this huge area, the ranger we met in Bettles was kind, helpful, and offered to loan us bear barrels—but we carry bear sacks, which are much lighter.

Our plan, after another few days on foot and in the packrafts, is to meet two friends for a resupply and a week's paddle together. We'll probably

1   In spring 2024, the Biden administration ruled against a proposed 211-mile road that would have cut through the preserve of Gates of the Arctic National Park just south of Walker Lake to access the Ambler mining area. The road would have potentially allowed more people to access the park, affected the Western Arctic Caribou Herd migration, and disrupted wilderness in general. Still, future administrations could repeal the ruling.

see other boaters, too, on the popular Noatak River. For the thousand or fewer visitors who actually set foot or paddle in the remote park each year, solitude is the name of the game unless you float one of the park's wild and scenic rivers (the Alatna, the Kobuk, the North Fork of the Koyukuk, the John, the Tinayguk, or the Noatak).

After oatmeal laced with nuts, we pull down the tent, shake off the frost, pack it away, and tramp a quarter mile uphill to the pass. I wear all the clothes (except the orange-stained, pepper-sprayed shirt and spicy ball cap) that I can find in my pack. Chris dons his down jacket with a look of singular pleasure at his preparedness. After the heat of last summer, along with this summer's wildfires throughout Alaska, it's not the weather I had expected, but neither of us would complain about a bit of winter over voracious mosquitoes.

My paddle is broken down and slotted in on either side of my pack. Chris has devotedly carried his Werner carbon-fiber paddle fully assembled in his hands for the last few days through tight alders, on river fords, and across scree fields. As if a creek or waterfall might suddenly flow our way and allow him to accelerate the pace. But it makes sense: in the past I used to hold my prized, sixty-centimeter, bamboo Chouinard Equipment ice ax in hand rather than on pack when I crossed tundra beneath mountains that I wanted to climb. Best to be prepared and to keep your hands busy in anticipation of the challenges to come.

I first met Chris two years ago. He had been assigned to photograph my magazine story about a sail across the Everglades, America's second-most-climate-endangered national park (after Gates). The three people in our party brought along rigid-hulled sea kayaks equipped with leeboards, masts, twin sails, and outriggers. But as a white-water technician, Chris didn't own a sea kayak—and since he came on short notice—we could only find him a wide-beamed, canvas-and-rubber Klepper kayak with a wooden paddle. But we had no sail for his tubby boat. So, for a week—while three of us blissfully rode the trade winds north through the swamp and along the Gulf Coast shores—Chris slogged along behind us. Through the binoculars you could see his double-blade work through the air like a metronome. Although it would have been impossible for a paddler to keep pace with sailors in such a wind, it didn't escape anyone's notice that Chris had to work his ass off while we just sat back and occasionally trimmed the sails or tapped on the rudders. We often waited for him, particularly if there wasn't a clear destination camp in mind. At the end of each day, an hour or two after we had unfolded our lawn chairs on deserted, white-sand beaches, Chris would pull in and go to work with his camera. He never complained.

We had also noticed that he'd blanched when we pointed out a toothy American crocodile that sunned itself on a beach alongside some alligators—probably harmless unless provoked. Characteristically, Chris said nothing, but you could see that something bothered him. In fact, as the quintessential professional, Chris said little the whole trip, but he was good-mannered, friendly, and, as the magazine editor had described him to me, "an old soul." Plus, he shot stellar photographs.

After the trip I figured out where I had heard Chris's name. Ten years earlier, in the Congo, on the previously unkayaked Lukuga River that drains the massive Lake Tanganyika, he and Ben Stookesberry accompanied the South African paddler savant, Hendri Coetzee. As the leader of their trip, Hendri had warned them to "stay off the banks because the crocs are having a bake and might fancy you for lunch."

Shortly after, just to the side and through his peripheral vision, Chris watched a fifteen-foot crocodile grab Hendri by the shoulder and violently flip him and his kayak upside down. For twenty seconds Chris watched the kayak hull quiver as the crocodile shook and wrenched Hendri out—never to be seen again. Chris and Ben went to the nearest village to begin a search, but the locals no longer had any boats because man-eater crocs had repeatedly pulled their friends and family out of the boats and eaten them. Chris vowed—out of respect for

Alongside Noatak headwaters too shallow to paddle, Chris passes through the wildflower arboretum of the Arctic, with (from left to right) *boykinia* (bear flower), cinquefoil, and dwarf fireweed (river beauty). JON WATERMAN

Hendri—never to try the river again. More than a decade later, he still grieves for his friend.

While Chris was traumatized, the experience didn't keep him off rivers. From the Arctic to Patagonia to Indonesia to South America, from vertical waterfalls to glaciated streams to jungle rivers, Chris would log more than one-hundred first descents in thirty-six countries. Camera always at hand.

Ahead of us, the Noatak is merely a class II scenic river compared to the class VI unruly torrents Chris has feathered his paddle through. So, his challenges on our trip will have little to do with the technical aspects of the river. For starters, as a polite and fastidious wilderness traveler, who regularly bathes and shaves despite the weather and lack of facilities, he will have to endure the bohemian style and gaseousness of his partner who will share the tent every night for the next month (freeze-dried food has always challenged my digestive system).

Chris is an introvert, happiest to keep his own counsel. When pushed, he'll describe himself with a favorite bumper sticker: "I'm not anti-social. I'm pro solitude."

I have traveled a similar behavioral geography. Witness my 2,200-mile trip alone across the Arctic.

Unlike me, Chris is not wired to ask questions. Wary of extended conversations, he has repeatedly been subjected to my monologues (as I am subjected to his silences) about the Philip Roth novels that I read at night. Then on the trail, whenever I can catch him, he has had to field (and sometimes answer) the stream of essay questions I throw at him in an attempt to get to know him and to slow him down.

Chris needs time to himself. Rather than ask me to stop and wait and then pose me, as most outdoor photographers would do, he'll anticipate a vantage point in the distance. Then without a word he'll hurry over and set up and compose a scene with me in it just as I catch up—no wait involved as I stagger past.

Shots taken, he'll slot his camera back in the huge dry bag of photo equipment he gently lifts back over his neck as his other hand clutches the paddle. The two clutched encumbrances balance the fifteen-plus pounds of bulky, rolled-up packraft and PFD that bulge off the rear of his huge pack. Then a long-legged sprint (at portage speeds that exasperate his white-water buddies) allows him to catch up to me and, just before I can throw out a third essay question, he blasts past.

Videography and photography will be his main challenge. On this trip he has sponsorship obligations that allow him to trade images and video for expenses and gear. Salaried sponsorship as a professional athlete has enabled Chris to spend much of his adult life on wild rivers.

I, too, have always been grateful for my sponsorships. But I remind myself that our principal mission on the Noatak is to try to understand and document—through video, still photographs, interviews, and journal notes—how the region has been affected by the climate crisis.

The drainage that we're about to walk out of is a classic example. Researchers have shown how the active top layer of permafrost (that refreezes in winter) that surrounds Kalulutok Creek has warmed and thickened over the last seven decades. Scientists predict that from the 1950s to the 2050s the average air temperatures will increase by more than 7 degrees Fahrenheit. Across the Arctic this means that thermokarsts will increase, more lakes will disappear (which will cloud the rivers with silt), and shrubs will eventually fur the tundra that we walk across.

The broad, knolled pass is a wonderland of wildflowers, isolated ankle-high grass, and a luxuriant emerald-green mat of tundra plants patched amid the rockiness of the alpine. Tiny pink moss campion, woolly lousewort, and purple saxifrage hunker out of the wind in their

Happy to get the heavy packs off our backs, we brewed up coffee and took shelter from the cold wind behind the biggest quartzite erratic I'd ever seen. CHRIS KORBULIC

own shrunken universe. Somehow these flowers absorb enough droplets of mist, melted snow, or thawed permafrost to survive—and to add color and beauty—in this polar desert.

Sheltered behind rocks out of the wind are stalks of two-foot-high, showy white bear flowers (a.k.a. Alaska *Boykinia*) with reddened stamens, amid a bouquet of azure-colored harebells. But now I have so far forgotten myself with flower photography that my fingers turn numb and I begin to shiver in the frozen wind.

Beyond the microview, the panorama continues to freeze me in my tracks. Immediately above us red, brown, and ivory boulders are as cleanly delineated as fingers that rake the steely gray mountainsides as if to scratch their torsos. Even though it's midsummer, the gullies up high are striped with snow like a zebra's neck.

"It's so beautiful up here," Chris says. "I don't want to leave." Easy for him to say: he's got a down jacket on.

Across the Noatak Valley, five miles west, the broad limestone-and-granite turreted Mount Igikpak—8,276 feet tall—stabs and pins in place a thick shawl of cloud (that will adorn its crown for days). Small snowfields and blue-gilled glaciers cling to a dozen other slender, jagged mountains that appear below the rush of cumulus.

Chris has his tripod out and works these peaks as if they're supermodels, while clouds throw shadows across strobed sunlight. He stands with his face glued to the camera under the hood of his down jacket. I run in place and windmill my arms to stay warm.

We ogle the famed source below: The nascent Noatak River threads through a panoply of boulders that would defy even the hardiest grayling. It appears to spring right out of the ground beneath Igikpak as glacial and snowfield meltwater. I know that Chris can't help but salivate at the chance to inflate the packraft strapped to his back and plunge into his element, but the steep stream more than a mile below looks too thin to paddle. Chris, nevertheless, will surely suggest that we try it as soon as we're down.

From here the river runs 65 miles to the western border of the park, another 265 miles through Noatak National Preserve, as it passes Noatak Village and meanders another 110 miles to its delta in Kotzebue Sound. As part of its 1980 designation and protection as a wild and scenic river, it remains the nation's largest mountain-river basin unaffected by human activity, dams, or development—unless you factor in the thaw that has already begun to radically change the North.

But now it's high time we thawed ourselves, shouldered our packs, and wiggled our legs downhill. We have lingered in the wind for over two hours on this pass. We're both excited to reach what appears to be a white wall tent, pitched several miles down valley, conspicuously contrasted against the slate-bouldered river and lawn-green velveteen appearance of the banks—even though we can't see any people. (I curse myself yet again: I left out the binoculars to lighten my pack.)

We follow a moraine field downhill for a thousand feet. After an hour of hops, skips, and jumps between boulders, my knees protest with crepitus as audible as a mouthful of chewed carrot.

We reach more level yet swampy ground in waist-high willows and pass shed caribou antlers until we emerge at the shallow river. Minnows dart through pools. Rosy fireweed beards the banks, splotched flaxen with cinquefoil.

The west wind whips and whooshes the willows sideways, as the distant and icy atmosphere of the Chukchi Sea reddens our faces. The stream of water purrs and shoals through gravel with equal abrasiveness against our eardrums.

Chris looks hopefully at the water, his paddle held as if it's a longsword. "Too soon," I say, "too bony."

Two ravens on some urgent matter of business ply the wind directly overhead. You can hear air luff through their secondaries like footsteps up an old staircase as they gain altitude to begin a long and speedy swoop. I can imagine, as the brainiest birds of the Arctic, that they're entertained by the two humans who are clearly out of their element and who stumble and limp with obvious overuse injuries, replaced joints, and weary agedness.

From the pass, we had fantasized that the wall-tent inhabitants would invite us in for a spot of tea out of the wind. "Probably park researchers," I had said to Chris. But from a quarter mile away it doesn't match our expectations. "It looks," I say now to Chris, "like a giant snowball avalanched down off the mountain."

We're still both puzzled, albeit disappointed, so we speed up our pace. From a short distance away, we realize that what we believed to be a wall tent is actually a fifteen-foot-high, ten-foot-wide white, glossy quartzite boulder with a reddened side of hematite that we perceived from the distance as a door. We've never seen such a massive chunk of quartzite.

Formed 150 million years ago miles below us as the tectonic plates began their high-pressure, super-heated rugby match, the Earth's crust folded and thickened, and forced the Brooks Range to slowly rise up from the sea. In the course of the uplift, over millions more years, the heat and pressure transformed subterranean sandstone—made up of quartz cemented together in round grains of crystals—into quartzite. Sandstone, quartzite, limestone, granite, and marine creatures fused into these rocks that all rose into the sky to become mountains.

The mountains aged. Glaciation took over. More than a hundred thousand years ago conjoined alpine glaciers swept throughout the Brooks Range and scoured out the valley that we stand in. A several-thousand-foot-thick glacier—rock-hard and able to exert huge pressure through its movement and sheer tonnage—plucked this quartzite protrusion from a mountainside and spirited it away. Over the millennia the boulder floated along, embraced inside congealed ice like krill that wafts below the sea surface.

Then, 16,500 years ago, as the local climate began to warm along with a measurable increase in atmospheric carbon dioxide, the glacier began to melt and recede. About fifteen thousand years[2] before our packs thudded to the ground beside the boulder, the once-massive glacier in the upper Noatak dropped its payload.

As the ice melted back, tusked woolly mammoths slightly bigger than the boulder stomped by, grazed in the meadows, and likely scratched their matted flanks against the quartzite. Back then, the headwaters roared with a deluge of glacier water in a natural cycle of warmth that the Brooks Range hadn't experienced for hundreds of thousands of years. Followed by a much larger, human-caused release of atmospheric carbon dioxide that has now changed the world.

Our arrival at this glassy landmark of antiquity is yet another acceptable excuse to stop for coffee. As the water comes to a boil, we palm the lustrous boulder and swat at a few laggardly mosquitoes that also take shelter in its lee, the wind effectively cut in two by one of the hardest rocks anywhere. Iñupiat used it for ulu blades. Below the Arctic Circle, it's now trendier than granite for countertops.

After a few desultory comments about the weather, the caffeine inspires us to walk for another two hours. Finally, under a sheltered bluff, we throw up the tent on damp yet soft sphagnum moss as comfortable as our Therm-a-Rests. We inhale freeze-dried green curry and stack the pots atop our food in bear sacks. Although the bulletproof Ursack fabric should keep the food out of hungry intruders' mouths, they might still play piñata with the bear sacks and reduce our food to powder.

We jump into our mosquito shelter. Chris takes out a hard rubber ball and puts it underneath his lower back—injured from his chainsaw mill work with huge cedars destined for his Olympic rainforest home

2 In April 2015, geologists from several different universities published their research on rapid deglaciation of the Brooks Range caused by an increase in atmospheric carbon dioxide 16,500 years ago. Through the measurement of the element beryllium in two erratic boulders in the Arrigetch—adjacent to the Noatak—they discovered that the boulders had dropped out of the receded glacier ice and were exposed to air about 15,000 years ago.

alongside his Airstream trailer. As I write in my journal, Chris rolls and stretches on the ball in hopes of pain relief. Since my right knee is now swollen, I take yet another dose of prescription anti-inflammatories so I can sleep.

Still, I note in my journal: *I am ever so happy.* I have schemed of a trip down the entire river's length since I first left in 1983.

After yet another oatmeal breakfast, I boil my ball cap with two changes of water in the pot. But an hour after we leave camp, my sweat activates the bear spray enmeshed in the hat and burns my forehead. So, I doff the hat and squint into the sun, frustrated that there's nothing I can do about the pants that similarly inflame my thighs.

Chris repeatedly ogles the river. I repeatedly shake my head.

We follow an oft-used bear trail, with the hind paws and forepaws of numerous bears perfectly embossed in the dried-out moss, eight to ten inches apart. As if navigating some sort of ancestral minefield, no bear deviates even an inch from the old pawprints. If I lifted and placed a ten-foot length of bear paw-signed moss on my bedroom wall—beside all my bird photographs and the wolf lithograph—the moss mural, too, would serve as timeless art. Or so I like to fantasize as an escape from the pack-laden slog.

I check for signs of meat eaters. Only one bear scat, whitened with caribou fur and dried out by the sun, contains crunched-up bones. Aside from the ravens, we haven't yet sighted any live animals, but I sense that we are watched as we clomp downstream.

Finally, at a grassy bench, the first Arctic ground squirrels we've seen or heard this summer chirp out their alarm calls as we come into view. Now that we're in the tundra region of the Noatak, instead of the forests on the south side of the Brooks Range, we'll be in the land of the adorable ground squirrels almost every day. Several stand up on their hind legs to peer at us from a distance before they duck into their burrows.

As the master of hibernation, the prairie-dog-sized ground squirrels—that weigh up to three pounds—continue to bark out their high-pitched *sik-sik!* (their Inuktitut name). In another two months, they'll decamp from the surface to spend the winter several feet down in their burrows. From September through April, their body temperature—unlike any other warm-blooded creature on the planet—will drop as low as 26.8 degrees Fahrenheit. In this supercooled state, their blood doesn't freeze as their brain neurons shrink and the neural connections shrivel. Every two to three weeks, to recover from their phenomenal torpors, the half-asleep squirrels will shiver until their temperature rises to 98 degrees Fahrenheit, and their brains undergo growth spurts and increase neural links beyond what existed before their hibernation. After little more than a day's shiver, with no food, they resume their winter life as frozen-pop squirrels.[3]

The abundant ground squirrels—named *Spermophilus parryii*, after their mushroom diet and the nineteenth-century explorer (of the Northwest Passage) Sir Edward Parry—are sought after by owls, foxes, wolves, and bears. As if they weren't already widely picked-upon prey, now researchers—alarmed about permafrost thaw—have identified ground squirrels as a small part of the problem. Because they mix up the soil when they dig their burrows, they bring in oxygen and then fertilize the ground with urine and feces. This, in turn, thaws the permafrost around their colonies and allows microbes to unleash frozen carbon and methane as greenhouse gases into the atmosphere.

No one yet understands how the complex and calibrated hibernation cycle of the ground squirrel will be altered by the early springs and late winters brought on by the climate crisis. Or how its diet of mushrooms, roots, seeds, flowers, grasses, sedges, and bird eggs will be jeopardized when the Arctic grows over with shrubs by the end of the century. The state of Alaska lists the species as vulnerable to climate change, with their population at moderate risk of extinction.

3 On a 2006 visit to the Toolik Field Station north of Gates, I learned about the resilience and wondrous hibernation of Arctic ground squirrels from Professor Brian Barnes, as he gently handled one of his newly captured subjects. Barnes implanted temperature sensors in a dozen ground squirrel abdomens while they hibernated in cages—maintained at 24.3 degrees Fahrenheit—several feet underground on the University of Alaska Fairbanks campus.

Chris, the whitewater ninja, happy to have ditched his pack inside the packraft amid the bumper car outwash of Tupik Creek, Noatak River headwaters. JON WATERMAN

Below the ground squirrel colony, as the headwaters swell, cliffs force us back and forth across the stream-almost-turned river. We jump from rock to rock and sometimes miss with soaked-foot splashes accompanied by curses.

In early afternoon, at one series of braided channels, a subadult wolf—perhaps sixty pounds—crosses the river in our tracks. Like a domestic dog, it comes closer to check us out. At a waist-deep pool, the lean wolf springs with feline agility five feet across to the bank. Black with a pale, buff band around the chest, unhurried, it vanishes into waist-high willows. Entirely camera-shy, it still makes my day—the young wolf had probably never seen people before.

Around the next bend at a big confluence, the stream swells into a small river to make Chris's day. As he stops to blow up his boat, I sprint downstream for another mile on a heavily tracked grizzly trail—and sing out, "Hey, bear!"—until the braids consolidate into a single river mostly free of rocks. When Chris catches up, his raft somehow intact after all the boulders and shoals, he pulls over to brew up.

The afternoon coffee has a kick, we discover, from the bear spray that congealed in the pot from the morning hat boil. Since we've never had capsaicin coffee before, it gives us a good laugh and an even better jolt. We drink every drop.

I gleefully ditch my Hyperlite pack and all its contents inside the inflatable tubes of the Alpacka raft and zip it up. To my surprise, the newly engineered raft (which I've never used before) could fit at least two packs with gear in its innards. I screw a threaded nylon sack into the boat valve, and then swish the other open end of the sack through the air to fill it up into a big balloon. Then I fold it shut, clutch the balloon between my chest and arms, and push the air into the raft. Like a blacksmith with a bellows (and pointers from Chris), I perform a dozen swished fills and pushes to inflate my raft.

We jump in, cinch spray skirts up to our armpits, and we're off. Pushed by the current, we make lazy paddle strokes and stare in wonder at the ancient limestone ramparts of the Brooks Range, scraped clean by glaciers, inside a narrow river valley that rises thousands of feet into a sky freighted with dark clouds. A rain squall gently hazes us, as if the sky had set its shower-stall nozzle at the finest mist position. We're caffeinated, rain-jacketed, and fastened snug and dry into our bulbous, mini-me rafts that turn and spin and stop on a dime.

For the next several hundred miles the pack-mule toil will remain behind us. One more day under heavy loads in trailless terrain, we agree—in shouts over the gabble of the river—would surely have damaged our bodies. As if it hasn't already.

———————————

At six in the morning, I wake up and clamber out of the tent onto the outwash of Tupik Creek. I pad barefoot across pale ivory quartzite sand, ground up by an ancient glacier and in the process of its return into the Earth. Along the river, a pair of northern shrikes chase each other through the dwarf birch in a silent and precise acrobatic ballet of dips, dives, and swerves.

In the brush I find an intact set of caribou calf leg bones—from the humerus to the ulna, radius, carpus and metacarpus, and foot phalanges. They are the size of my arm. The species is born to run, and at two days old, this calf could've outdistanced me. An adult can run at over forty miles an hour. The tuktu leg is all still attached to tawny bands of cartilage. I hold the bones up and pivot the assemblage back and forth to visualize the anatomical wonder that propels the species thousands of miles each year.

From the cool shadows I watch the sun lay a diagonal band of gold across a distant, verdant mountainside. Washed by yesterday's rain showers, the air is transparent and allows the light to pass over the land and illuminate it with the resolution of a polished gem. I am transported by the thought of wildflowers that open their blossoms under the dawn light in the age-old choreography of photosynthesis.

Yet again, I am haunted by the Arctic light.

As it comes down and warms camp, we stretch on the sand, bathe in the river, and wander like naked Neanderthals. I dry off in the sun.

Alongside the tent in a huge patch of fireweed I crouch in the shade of thin, shrubby cottonwood trees to photograph scores of Arctic bumblebees. Thanks to their pollination, there are flowers everywhere in the Noatak valley.

Arctic bumblebees have made extraordinary adaptations to survive in the North over their forty-million-year evolution from wasps to bumblebees. Compared to the Lower 48 bumblebees, the Arctic bumblebees are significantly hairier, clad in yellow, fuzzy jackets. They're also huge. To warm up they perform a long muscular shiver as they vibrate the pollen off the pink-tipped, long-armed stamens of the fireweed. The bees, in turn, are fueled by this same pollen. They hang—dusted with white particles—and spin from the lengths of stamens like silk dancers.

Unlike their warm-blooded photographer in the shade, the bumblebees linger in the reflector flower petals that magnify the heat of the sun. When I close my eyes, the buzz and vibratory tremor of these bees next to my ears overrides the gravelly purl of the nearby river.

Everything changes when the throaty roar of a bush plane—with our friends Gary and Carolyn—circles downstream, then heads out of sight a dozen miles away to land on a gravel bar. We break camp quickly and hit the river with our paddles, excited for company, resupply, and the promise of roasted vegetable and chicken fajitas.

Picking through loose scree with Gary, Carolyn, and Jenny (the dog) up to Peak 4880 above the Noatak headwaters, amid rain showers and distant wildfire smoke that would burn three million acres that summer. CHRIS KORBULIC

Wildfires

July 16, 2022

The five of us—Gary's dog, Jenny, included—leave camp the next morning to climb a peak several thousand feet above our river camp at 1,800 feet. All the way up the steep tundra slope, littered with scree, ground squirrels chirp amid a riot of blueberries. We pop them into our mouths for bursts of sweet-tartness, but quickly, so that Carolyn—in thick-soled Hoka boots that seem to work as pistons—doesn't completely leave us in the dust. The bearberries are still hard and reddish-yellow astride their ovate green leaves. The black crowberries—that plumply bow over several-inch-long stems leafed like miniature fir trees—are also too firm to eat.

After three hours of breathless labor behind Carolyn, we clamber the last fifty feet across ancient limestone up a peak marked on the map as 4,880 feet high. My old friend Gary rigs a fixed line for the final section of loose rock, but Jenny, an experienced mountain climber, doesn't need it. We speak in reverential tones, huddled atop our narrow altar to contemplate the universe.

The river curls powder blue beneath our feet. We're mesmerized. The view is surreal, a landscape architect's scale model of Eden before the fall, with no sign of humanity. All dominated by the sinuous river.

Twelvemile Creek rushes in from the northeast to further fill the northwest flow of the Noatak. It occurs to me that—in combination with its

PREVIOUS SPREAD: Chris surveys the Noatak headwaters valley, increasingly overgrown with shrubs and wildfire smoke that made our eyes sting. Fortunately, we had few smoky days. JON WATERMAN

hundred-mile-distant twin to the south, the Kobuk River—these huge waterways are the source of life and virility to northwest Alaska. To get to know this huge place, Chris and I have deliberately and repeatedly scouted the region. We flew over the entire river in a small plane before we ever put on our packs. Then we trekked to its source. And today's climb to survey the Noatak yet again.

We can no longer hear it below us. The river is as dreamy as a thawed glacier, authoritative in its sculpture of the smoky valley.

Then there's the Noatak's contribution to biodiversity. Its fishery. And its natural beauty, never to be dammed or denigrated or parceled out by the acre-foot for hydroelectricity or agriculture. The river remains protected simply for its wild value to the world.

For a couple of minutes, Mount Igikpak emerges from the clouds through the smoky haze, four thousand feet higher and a dozen miles to the southeast, to finally show its pinnacled, cat-ear summit. By comparison, our airy perch feels pedestrian. I look away, look back, and the elusive Igikpak is gone.

The haze and caustic pong come from hundreds of wildfires that burn throughout the state. In the previous month, two million acres burned in Alaska. At the end of June, and into July when Chris and I arrived in smoky Fairbanks, the city and four nearby towns had the worst air quality in the United States, even though northern Alaska is known for its clean air.[1]

Over the past several decades, the national trend in these times of climate change, amid greater heat and extended drought, is toward larger and more frequent wildfires.[2] Wildfires have long been part of a natural and regenerative cycle, particularly in the boreal forest on the southern fringes of the Arctic. But recently fires have begun to blaze up into the Arctic. For the last eleven thousand years Arctic fires were extremely rare, but now the dried-out, warmed-up tundra with its flammable peat and newly grown shrubs provide fuel (recent climate change has also

[1] In the Fairbanks area, tundra fires created a health hazard fifty times over what is deemed acceptable by the World Health Organization (WHO). Even amid the sea breezes of coastal Nome—350 miles southwest of our smoky peak climb—fine smoke particulate matter rose to 140 times that deemed healthy by WHO. The smoke, which came from fires hundreds of miles away in Alaska, decreased visibility to under a mile in Nome.

[2] From 2001 to 2020, wildfires in Alaska burned 31.4 million acres, more than two-and-a-half times as many acres as burned from 1961 to 1980 or from 1981 to 2000. Statewide in 2022, three million acres burned, which is three times the average of the past decade and three times the amount of acreage that burned in all forty-nine states that year (albeit a low fire year in the Lower 48).

caused periods of extreme and unusual rain). Lightning, also previously rare in the Arctic, starts most of these new fires (in north-central Alaska from the mid-1980s through 2015, lightning strikes increased 600 percent). Longer summers have extended the fire season, too.

In addition, after the burned blanket of peat, brush, and tundra plants releases huge amounts of greenhouse gas, the exposed permafrost spews out more carbon—and methane. There is twice as much carbon in the world's permafrost as contained in the atmosphere.

In 2007, on the other side of the Brooks Range from where we sit, more than four hundred square miles burned around the treeless Anaktuvuk River. There has never been such a large fire so far north. Caused by a lightning strike, the fire released two-million-plus tons of carbon into the atmosphere—as much carbon as the world's tundra had stored over the past half century.

Ten years later, astonished by the extraordinary fire, University of Alaska researchers visited the burn scar to evaluate the long-term damages these new fires would inflict on the Arctic. The willows had regrown rapidly and were even taller than other newly shrub-infested areas of the Arctic. Thermokarsts slumped throughout. Other essential Arctic plants—such as the sphagnum moss and lichen that the caribou depend upon—had declined.

To the west of our aerie—like raptors in search of new terrain—we spy patches of blue sky where we might pass out of the smoke. But none of the Noatak is exempt from fire.

The frequency of wildfires in the landscape that surrounds us—particularly in the Noatak National Preserve downriver—has greatly increased over the decades. In the 1950s and 1960s, more than fifty thousand acres burned in Gates. In the previously fire-free preserve, nearly two hundred thousand acres burned in the 1970s. The fires would continue to increase over the next few decades in direct relation to increases in temperature, shrub growth, dried-out tundra, and lightning storms.

Clearly, I have too much information in my head[3]—if you're not careful, this climate crisis evidence can ruin your day.

Clouds and smoke continue to build around us as we clamber down to camp. On the steep scree and tundra descent, my knees become castanets. *Good thing*, I think, *that we'll mostly be in boats now*.

As the group continues its descent, I deviate into a gully. At a band of steep limestone, a spring bubbles out of the ground and cascades into a waterfall. I follow a game trail through the cliffs and at a gin-clear pool lively with water striders, I crouch and make like a tuktu to fill my stomach.

At camp, Chris's measurement stick from the previous night shows that the river has dropped several feet. Gary and Carolyn pull out more fresh vegetables and cook for us yet again, sheltered from the wind behind their overturned canoe. Jenny cowers in the tent.

Before I can finish a catch-up conversation with Gary after dinner, the rain starts, and we sprint for our tents. It pours all night long.

3 I had recently finished two intensive research projects for National Geographic's atlases on Gates and the Noatak National Preserve with help from the National Park Service. The fire data and climate effects for both places are unforgettable.

Downriver
July 17–21, 2022

1   In 2019, the *Guardian*, a widely read British daily newspaper, changed its style guide to use the terms *climate emergency* or *climate crisis* instead of *climate change* to more "accurately reflect the seriousness of the overall situation." When it's appropriate to describe something like "global warming," the *Guardian* now substitutes "global heating." Instead of "climate sceptic" they use "climate science denier" or "climate denier." Liberal and conservative papers in the United States—that include the *New York Times* and the *Wall Street Journal*—still use the terms *global warming* and *climate change* (but rarely *climate crisis*) in their news stories.

I first bumped into Gary Kofinas in the summer of 1988 in the village of Kaktovik. He had come to Alaska to climb mountains and, later, to help launch a new University of Alaska graduate program in sustainability. There in the typical Barter Island pea soup fog that grounded planes, Gary, a second-generation Greek-American who grew up with the ancient language spoken in his home, had a curious, friendly manner that drew me in.

We exchanged addresses. The next summer he hired me to lead students through the Arctic on a college-accredited "Wilderness Versus Oil Development" course, through his Expeditions for Global Awareness nonprofit. Over the years we would log time together across tundra and down cold rivers.

Although a warmed-up Arctic wasn't on most people's minds in the late 1980s, Gary's doctoral studies delved into Indigenous communities and climate issues. The soon-to-be Dr. Kofinas couldn't help but suggest to everyone he knew to use the term *climate change* rather than *global warming*, because of the many environmental dimensions beyond temperature increases that had already altered the North.[1] Although scientists had begun to understand the magnitude of the changes, back then the public was still uneducated.

PREVIOUS SPREAD: As we continued downriver through different ecosystems, we often passed a different and taller (up to three feet) genus of fireweed, known to colonize areas recently burned in wildfires. Iñupiat ancestral lands. CHRIS KORBULIC

During his field research Gary met Indigenous elders and hunters who told him how the land had changed, how they could no longer predict the weather, and, as migrations changed, how their subsistence hunts challenged them to no end. One Inuit from Canada said his friend Gary and other researchers were like *sik-siks* (ground squirrels) because they only came up in the summer, versus the northern villagers who lived, breathed, and suffered the changes firsthand. The seeds were sown for Gary's move to Fairbanks as a professor.

Through his village travels with the University of Alaska, his interactions and friendships with Indigenous peoples provided new insight into changes across the North and consistently showed the integrity of the villagers' experiences and awareness of climate changes on the land. These observations often incited scientists and wildlife officials to investigate and take action.

As professor emeritus of Resource Policy and Management, Gary continues his research at the University of Alaska's Institute of Arctic Biology. While his dozen-page CV lists a lifetime of academic accomplishments and publications (such as his contribution to reports for the Intergovernmental Panel on Climate Change), his profile photograph shows him clad in a Gore-Tex (rather than sport) jacket, tied to a rope with a huge grin, above a glacier (instead of a desk) high up on an Alaskan mountain. Late in his seventh decade—fitter than most men two decades younger, while forced to confront the deaths of two different life partners from cancer—he routinely migrates to the Arctic. He is more tuktu than sik-sik.

———————

By midmorning, Gary and Carolyn lead the way in a speedy canoe through a Noatak silted a fainter, cerulean blue. Braids split the river, and we repeatedly hop out of our packrafts or canoe and line the boats down shallow channels.

Rain spits, while mergansers scoot and grunt; *gruk, gruk, gruk!* They beat the water in arpeggio with pointed wings that pull them into air with a prolonged *whirrrrrrrr.*

At one sharp bend the river has exposed a several-foot lens of white permafrost ice, inside a small cave of black duff, topped with a scruffy hairdo of pale green grass. Chris and I linger and work our paddles to spin in an eddy as we breathe in a sharp, peculiar fennel redolence of ancient earth uncovered—the smell, perhaps, of microbes released in the thawed soil, as carbon and methane rise from the permafrost.

Within only a few miles the valley broadens. At first, it spreads a couple of miles wide, then it doubles, and afterward, distances are hard to discern. The river, no longer a wild mountain stream, runs mostly through a single channel, as busy with looped meanders as a roller coaster laid sideways.

We pull over at Pingo Lake (from last summer's put-in with Alistair). The popular floatplane entrance for Noatak paddlers now has a well-trafficked portage trail that had grown over in the low-tourism, pandemic years. A dozen paddlers at the lake prepare to float the river. With convivial wilderness-traveler etiquette, we trade notes and where-you-froms amid the sound of insects that get slapped on the necks of those adverse to head nets.

Pingo Lake is a vivid reflection mirror for the distant pingo, a volcano-shaped hill that rises seventy feet to the west. Pingos are born in drained lake basins or on slopes fed by underground springs. As water froze into a lens shape underground several centuries (or more) ago, the ice lens expanded with more water and forced the tundra up into a mound, and eventually, up into a hill.

To judge by the shaggy-headed, heavily vegetated pingo in front of us, pingos are refuges for plants, as well as prime habitat for bird nests, and at their breezy tops, caribou mosquito relief. Across an otherwise flat landscape, pingos (or *pinguq* in Iñupiaq—a term first borrowed by a

Canadian botanist in 1938) provided game lookouts for northern hunters long before the Europeans arrived.

Pingos feature prominently in an Inuit origin story from the Mackenzie River Delta about a time of ominous blackness in the sky. A hunter who paid attention to such environmental changes had his family build a raft atop a pingo so they could survive the flood that came and inundated the Earth and drowned all its inhabitants. From the pingo they set off across the waters. One day the hunter's son—inhabited by the spirit of a raven—set off from the raft in his kayak to explore. Just beneath the surface of the sea, he found a submerged island. He harpooned the pingo, and this caused it to rise and the floodwaters to recede. The event heralded a new beginning for the people of the delta, who flourished because they made new accords that respected the Earth.

Today, from the Mackenzie to the Noatak, as the world rejects climate protocols and accords, pingos have begun to collapse in greater numbers.[2] Warmer conditions melt the ice belowground and collapse pingos into craters (under past cold conditions, pingos collapsed when the ice lens below continued to push up and cracked open the earthen crust, which exposed the ice to open air). Born in the ice, pingos now routinely die in the thaw—sometimes explosively.[3]

As we reboard the boats and slap at mosquitoes, we can see a landscape filled with more permafrost features, where flatness is only an illusion. The smaller *palsa* hills are formed by large frost heaves in boggy ground. The elongated *flark* depressions are often filled with meltwater alongside distinct frost-heaved *strangs*.

Amid all the strange landforms, Carolyn—on her first visit to the Arctic—can't help but feel intimidated. Still, as an athlete and a camper who regularly competes in cross-country ski, mountain bike, and ultrarun races, Carolyn Murray isn't shrinking from the challenges of the Noatak. Her experience, with fresh perspective, speaks to the unique yet bizarre aspects of the Arctic.

[2] A 2020 study performed by a team of international scientists found that continued levels of greenhouse gas emissions would cause a loss of about a third of all pingos by 2050; between 2061 and 2080, more than half of the pingos are expected to collapse. About eleven thousand pingos can now be found across the globe.

[3] On the Yamal Peninsula in Siberia, a few degrees latitude north of the Noatak River headwaters, pingos subjected to rain and thaw (and underlain by gas) have recently begun to explode and throw earth and ice hundreds of feet into the air. The pingo craters that remain—at least seventeen have been counted on the peninsula—fill with water up to 160 feet deep. A collapsed pingo is often called an *ognip* (pingo spelled backward).

Chris, Carolyn, and Gary seek warmth from a cold wind, partially blocked by the overturned canoe, near the western boundary of Gates of the Arctic National Park. JON WATERMAN

It started with the mosquitoes. As we push off from the bank, I mention how mild the bugs are. Gary and Chris smile with tacit acknowledgment at my remark and the psychological survival tools that help us overlook and cope with the mosquitoes: if we say the bugs are mild, maybe we can believe it. Carolyn rolls her eyes. A good sport, she says nothing about the swarm that surrounds us.

Add to this the continuous daylight. The first couple of nights—in the exalted land of the midnight sun—had excited her. But now she feels unsettled and uncomfortable (as she later confided to me).

One night, she pulled her eyepatch up to see if it was time to get up. The tent, of course, was bright as day, even though it must've been midnight; she felt as though she was no longer allowed to sleep. Her circadian rhythms were off. Meanwhile, Gary sawed logs in dreamland. Describing her anxiety, she told me it was "as if my body has gone outside of itself."

Like most first-time Alaskan campers, the fear of a grizzly entering our camp at night preys on her. Before the trip, Gary had plunged into full disclosure mode—his patent honesty unimpaired by discretion— and told Carolyn a bear story. It involved a couple he'd met on their way into the Arctic wilderness who shortly thereafter had a tragic encounter with an aggressive grizzly. If not for the shotgun Gary placed at their side at night, she wouldn't have slept at all.[4]

The trip delights Carolyn, too: our slow movement downriver, not bound to any particular timetable or destination camps other than the prearranged, end-of-the-week floatplane takeout at Kavachurak Lake. This night, as kids in a candy shop, Carolyn and Chris hunt for rocks of all colors and shapes and forms—spewed up from the uplift of the Brooks Range over the ages and washed down along the riverbanks. And Carolyn enjoys the routines of camp setup and disassembly, meal preparations, and their canoe pack each morning.

But it doesn't escape her notice that the five-year-old dog Jenny is out of her element. Normally an outdoor dog, independent, and confident,

4 In 2005, an experienced Alaskan couple took a river trip in the Arctic Refuge. Every night, in deference to and with caution about barren ground grizzlies, they cooked away from camp and kept all toothpaste and smellables out of the tent. Even though they slept with a high-caliber bear rifle between them, it wasn't enough to stop the grizzly that collapsed the tent, killed them, and began to consume them. (Carolyn didn't hear the part about the gun in the tent. Nor did she know that for a couple of nights on the Noatak, the gun dry bag in her tent was empty because I had the shotgun in my tent.) The grizzly was later shot and killed, but the necropsy did not reveal any abnormalities.

bred from a member of a champion Alaskan sled dog team, Jenny spends most of her time in the tent. In the morning, yet again, they have to drag her out and place her in the canoe. Jenny's new timidity feeds Carolyn's apprehensions—and, perhaps, vice versa.

On July 18, the temperature drops, and the winds pick up. The sky darkens. And our plan to climb the glaciated Oyukak Mountain is squelched by rain and snow that dusts the terrain immediately above us as white as the powdered sugar on a Bundt cake.

As our teeth chatter in the cold, Gary reminds us that "while the overall trend is warming, that doesn't necessarily capture what's going on [with climate change]. Because the emergent pattern is increased variability."

I ask him to explain what causes the variability.

"It's important to remember that warmer temps put more energy into the Earth's system," he replies, "resulting in a greater variability of conditions. As a result, we're experiencing greater extreme climate events and fewer average days, which in turn acts as an amplifying feedback to the system."

The cold weather, however, is not all bad news. After eight days of constant travel, my whole body aches, and a day in the tent to read is a huge relief, so I return to the imaginary urban world of Philip Roth. Between squalls, we wander out through bogs and strangs and flarks, around palsas. A duck egg lies unattended on bare tundra with no nest, but the hen undoubtedly lurks somewhere nearby. We collect dead willow trunks for a fire—in the 1980s I hadn't seen such thick and prolific dead wood in the headwaters.

We soldier on through that day, then paddle on through the next. We're briefly trailed by a red fox that ducks in and out of the brush and keeps pace with our boats, undoubtedly curious about the blonde canid that ducks and shivers in the canoe like Jack London's Buck about to be attacked by a wolf pack.

We spy a prodigious rack shake through the willows above a curious moose, that outweighs all of us put together, as he breaks more branches to get closer. The narrow strip of cream-colored fur atop his back matches his rack. When he sees that we aren't moose, he dashes away in horror, followed by the sound of more splintered willows—the proverbial bull in a china shop.

As the rain continues, the swollen, once-blue river dulls into iridescent flint. Silt pumps up and down in the current, spasmodically surges to the dark surface, and for scant seconds, blooms open as pale white, half-dollar-sized Rorschach inkblots that invert like jackets turned inside out. The Brooks Range, arisen from the sea, now ever so slowly, migrates back to it.

Wind pummels us in the oxbows. For a mile, as the river turns northeast, we fly downwind in minutes. For the next mile, we point west into the wind, put our heads down, squinty-eyed, and splash-punch paddles until our arms ache and our hands go numb. For nearly a half hour we make precious little progress through brown blizzards of silt lifted off the riverbanks that sweep upstream and rake the Noatak into wavy, reverse-whitewater runs. Two missed paddle strokes and the wind blows you back upstream. This is the Noatak of river-runners' legend.

At the next bend we turn downwind with relief and wear the wind braced against our backs as it scoots us downstream as fast as we can paddle. Pale green-leafed willows on the bank bend in subjugation and the leaves wave their white fuzzy undersides like flags of surrender to the wind.

When we turn a corner in the roller coaster for another battle—back into the brown howl of a river-become-dust-bowl—I suggest a break. We pull the boats ashore, grab some dry bags, and duck into the lee of the tall willows to share chocolate and bread that we smear with peanut butter. From underneath two jacket hoods, Carolyn nods toward the river and says, "My enemy rather than my friend."

Often in pain from a hamstring injury aggravated by long pulls in the canoe that forced her to sit or kneel, Carolyn distracts herself with thoughts about how humankind's need to conquer and develop nature would be thwarted in the Noatak. Even without its national park protections she believes that the place is simply so unfriendly and wild that humans could never control it.

She also wonders how anyone could call such a place their home. As a mother who has raised children and traveled the world and seen other cultures, she still can't imagine how the Iñupiat can feel comfortable amid such stretched-out, treeless horizons, with summers that suddenly turn to winters and winds that sandpaper your face.[5]

The Noatak Valley has long been an inhabited wilderness—contrary to most modern Americans' conceptualization of national parks and public lands as places where humans are only visitors. Most of the continent was inhabited prior to the colonies. Native Americans didn't think of it as—or call it—wilderness because it was simply their home.

After statehood in 1959, Indigenous Alaskans wanted to do something different with their homeland. The controversial plan got a push from several Indigenous representatives, such as the Iñupiat Willie Hensley, who grew up on the Noatak River Delta, where his family subsisted off the riches of the land and its wildlife.

His 1966 college paper "What Rights to Land Have the Alaska Natives: The Primary Question" helped galvanize a statewide movement and unite the Indigenous peoples into a political force to be reckoned with. Hensley and other Alaskan Natives employed the power of public opinion and legal jurisprudence against the miners, the business community, the loggers, the state government, and the oil industry that sought to drill and build pipelines across the state. In 1968 massive oil reservoirs had been discovered in Prudhoe Bay.

5 Carolyn Murray experienced her most profound realization after she got home. As a schoolgirl, she had been inculcated with the clichéd stories about Eskimos who lived in igloos (structures that were only built when Iñupiat traveled in winter; otherwise, they lived in sod or whalebone-and-driftwood homes). Carolyn was a reporter with a Utah NPR affiliate station, and after her trip to the Noatak, she became aware of the many news stories about the Arctic climate crisis and how many northern villages were in upheaval.

She says: "We're so wrapped up in our own world that we don't appreciate other people of the North and their climate changes. Now when people ask me about Alaska, that's the first thing I go to: the people up there in a world of challenge. Now when I look at photographs of us, smiling, huddled behind the canoe, there's a sense of accomplishment. I did something that was pretty big."

As we enter Noatak National Preserve, we paddle beneath sand bluffs deposited by
glaciers during the last Ice Age that dammed up the river and created the (now drained)
1,700-square-mile Noatak Lake. CHRIS KORBULIC

Complex negotiations and unlikely alliances ensued. Ultimately, the state and the Indigenous population worked with the oil industry to secure land from the federal government. The state wanted to begin land leases to the oil industry; the Indigenous Alaskans wanted land so they could continue subsistence lifestyles, but also to profit from resource extraction on the land.

Hensley's leadership through the state house of representatives and his travels from Kotzebue to Washington, DC, to lobby for his peoples' rights came to fruition in 1971. That year, Congress quickly passed the Alaska Native Claims Settlement Act (ANCSA), and President Nixon signed it into law.

ANCSA equitably broke a lot more ground than the century-old purchase of the territory from the Russians or even the recent statehood act. The new law divvied up pieces of the pie for Natives, oil companies, and (eventually) environmentalists. Rather than schlep sixty thousand Alaskan Natives onto reservations—as the dysfunctional Bureau of Indian Affairs had done in the Lower 48—ANCSA gave the Iñupiat and the other original inhabitants of Alaska nearly a billion dollars (largely through taxation of the oil industry), more than forty-three million acres of land in twelve regions (a thirteenth would be added later) of the state. A dozen different Native Corporations own the land and work toward profitable ventures for their tribal share owners.[6]

The act intended to preserve their culture and keep villages intact. It also allowed the Iñupiat and other Indigenous cultures to find a small level of economic security and become part of the capitalistic collective.

At zero hour, DC lawmakers had added the subparagraph "d-2" into the long-winded ANCSA document.[7] The buried subparagraph would serve as the complementary lynchpin for a more massive landmark law that would eventually satisfy the neglected environmentalists. With the interests of wilderness lovers throughout the nation, the law would be called the Alaska National Interest Lands Conservation Act (ANILCA).

6   NANA (the Northwest Arctic Native Association) Regional Corporation has roughly fifteen thousand Iñupiat, or half the state's Iñupiat population, and encircles the region that surrounds the Noatak River. Natives who live in homesteads, villages, and the town of Kotzebue are all members or shareholders of the corporation.

Despite the best intentions of many Qallunaat legislators and Native advisors, today there are still some Indigenous people who are angry about the act and what they perceive as federal interference and assimilation of their culture. For example, an anonymous Gwich'in reader (whose perspective helped clarify Indigenous peoples' culture in this book) had this to say about ANCSA: "The act was assimilative and was meant to destroy tribes and force corporations to develop lands, forced into capitalism and not our traditional values."

7   The little-known d-2 clause within ANCSA directed the secretary of the Interior to prevent development on up to eighty million acres of land. Congress could then designate these d-2 lands to be made into national parks, wildlife refuges, wild and scenic rivers, or national forests. The rancorous debate that ensued would pit the adolescent state of Alaska against its older federal parents.

The decade-long haul to its creation involved many public councils and land explorations. It also poked a willow branch into a grizzly den of controversy: since it proposed to double the land allotments of ANCSA, ANILCA easily created twice the controversy.

Congress, in fact, proposed a lot more than eighty million acres of new conservation lands. Alaska saw their raise but bluffed with only twenty-five million acres. In November 1978, the House passed the bill, but the Senate became gridlocked as one of two powerful Alaskan senators threatened a filibuster.

In December, just before the so-called d-2 lands provision expired, the conservationist, human rights–oriented President Carter acted. Worried that oil and other development industries would ruin Alaska wildlands and shortchange Alaskan Natives, Carter used his executive privilege under the 1906 Antiquities Act to rescue and set aside fifty-six million acres of the d-2 lands as national monuments (which included Gates and the Noatak area). Emboldened by Carter, the House reproposed 127 million new acres of more widely protected parks and conservation lands. In Fairbanks, an angry mob burned Carter's effigy. In Denali National Park, protestors fired their guns (illegal then in national parks) and committed other aggressive acts.

Two years later, Alaska capitulated, and both houses of Congress finally reached a consensus, through many compromises (such as massive forest clear-cuts in southern Alaska). Carter signed the ANILCA bill in the last full month of his presidency. In Gates, insurrectionists severed a vital cable on a National Park Service plane—if it hadn't been discovered, the plane would've crashed.

ANILCA remains the single largest expansion of protected lands in US history; and the act more than doubled the size of the entire country's national park and wildlife refuge system. The law permanently protected 104 million acres.[8] The breadth of newly set-aside land staggered those geographically aware Alaskans, many of whom would remain up

8 There were already forty-seven million acres of national parks and national forests in Alaska. After ANILCA became law, the protected federal lands swelled to an enormous 41 percent of the state. As the largest state in the union, Alaska is more than twice the size of Texas, the second-largest state.

July snowstorms—once common in the Arctic—still occur as warmer temperatures put more energy into the Earth's atmosphere and cause greater variability and extreme weather events. Summer hardly came in 2022. CHRIS KORBULIC

in arms for years about their now "locked-up" land.[9] It was as if the peanut farmer from Georgia had carved an equivalent acreage—Maine, Vermont, Massachusetts, New Hampshire, Pennsylvania, New Jersey, and New York—off the Eastern Seaboard.

The newly created Arctic parklands included Gates, the Noatak National Preserve, the adjacent Kobuk National Park, Bering Land Bridge National Preserve, and Cape Krusenstern National Monument. Along with the creation of other parks throughout the state, Wrangell-St. Elias National Park and Preserve became the largest national park. The act also added wild and scenic rivers, wildlife refuges, national forests, and monuments, and doubled the size of the Arctic National Wildlife Refuge (formerly the Arctic National Wildlife Range).

ANILCA, with respect to land rights already granted to Natives state-wide through ANCSA, also provided for Iñupiat in its Arctic parklands. For Indigenous Americans in the Lower 48, parks are now little more than stolen land, and with few exceptions, they are closed to habitation or wildlife harvest by their original settlers. But here, in Gates and the other Alaskan parklands, the people from nearby villages are allowed to live in, hunt, fish, trap, and continue the subsistence lifestyle that they have practiced for millennia.

—————————

We stop for a quick brew alongside Lake Matcharak, with its seven thousand-year-old lithic and bone remains from ancient hunts. Like many other places in the parklands, this site (hidden somewhere along the lake) exemplifies the long tradition of the Iñupiat and the ASTt who subsisted on the land.

As I hunt for our stove, some bewildered first-time Arctic paddlers show up and pull off the river. I point out the route across the tundra to the lake and learn that they had originally planned to continue downriver for several days. But because of the unseasonable cold and wind, they used their satellite communication device to

9 In the early 1980s, I lived in a small Alaskan town and worked as a Denali National Park Mountaineering Ranger, tasked to advise or rescue climbers and conduct patrols on the mountain. My coworkers and I seldom wore our uniforms outside the ranger station because of the strong anti-park service, "stolen-lands" sentiment throughout the state.

arrange an earlier floatplane flight out from the lake. We can hardly blame them.

Below the lake, the land becomes a broad plain for a dozen miles. That night in camp the wind hammers our tents as the rain knocks to get in.

By morning, the river is a chocolate flood, several feet above Chris's daily measurement stick, alongside a canyon that squeezes the river into a roar and rachets up at least three paddlers' anxiety. We zip our dry suits up tight. The swollen river's eddies are washed out and the water is now opaque and hard to read.

High riverbanks shelter the bank swallows that twitter and peer from their darkened nest holes. They dive into the air and swoop, batlike, after unseen bugs—or maybe to practice aerobatics—above our heads. They cry out, *chir-chir-chir!*

The wind whips into our faces, but the current is so strong and the river so full that we are carried, exuberantly, in a breathless rush between narrow walls. Since I've been down this stretch of river before, I lead the way. Last year, I avoided a class II ledge on river left that would've held or flipped our boats. Yet today, the flood has turned this stretch of the Noatak into even more turbulent class III waters. I cautiously hug the cobbled right-hand shore.

After a mile, Gary speeds past me in the canoe and, over the pandemonium of water that rushes to the sea, loudly instructs Carolyn not to grab the gunnels—as Jenny cowers flat in front of Gary in the stern. To my horror, they brazenly steer river left directly toward the ledge. Luckily, since the flood has mostly washed it out, they blast through the drop with a quick plunge that nearly submerges their bow, but their momentum allows them to wetly punch through a wall of water.

Dr. Kofinas lets loose a whoop. Ms. Murray remains silent in the bow.

Thermokarst Landslide
July 22, 2022

Cotton-ball cumuli sail through blue skies and clash with darkened cumulonimbus warships that ominously obscure the horizon. The rain comes down in sheets that could've sent reverent, olden-day hunters to pingo tops. But the squalls pass quickly.

We ride the river's semicircles out of Gates and into the Noatak National Preserve. In the contiguous United States, we would have at least passed a sign, groups of boaters, or maybe a ranger station, but here only an uninterrupted and wondrous plain of greenery rolls off to distant, snow-frosted mountains on either side of the river. Less regulated than its national park neighbor, created through ANILCA with Iñupiat as much in mind as caribou and wilderness, the preserve before us also encompasses an age-old way of life. The six-and-a-half-million-acre preserve—the size of Massachusetts—continues to provide food security and a place for the Iñupiat to continue traditions they have practiced for thousands of years. Otherwise, Noatak Village has a grocery store with incredibly overpriced and understocked food supplies.

Farther downriver, we expect to see Iñupiat hunters and subsistence cabins. This high up in the preserve, more than two hundred miles from Noatak Village, across innumerable shoals, against the current,

PREVIOUS SPREAD: As permafrost thaws it causes thermokarsts, exposing more permafrost to warming air as the melt increases and sluices a channel of slow-moving mud and water down to the Noatak River. CHRIS KORBULIC

it would be too time-consumptive and expensive to travel here by boat. Gas costs villagers eighteen dollars per gallon.

Several miles from Gary and Carolyn's takeout at Lake Kavachurak, we come to a huge landslide. The thermokarst is unrecognizable from last year's ten-foot-high cave that looked on the verge of collapse. This year the landslide has subsumed the cave with mud and created a delta that forces the river to flow around it. Above, the thermokarst has grown into a huge, four-hundred-foot-high, three-hundred-foot-plus-wide amphitheater of slurry canals. Rocks periodically melt out of the amphitheater and skitter down dirty, metallic-sheened ice walls to splash into a gray silt pond the consistency of runny cement.

Thermokarsts and permafrost thaw are not new to the Arctic. About fourteen thousand years ago as glaciers began a rapid thaw (discovered by University of Alaska researchers), the runoff overwhelmed the capacity of stream systems to transport the mud and soil down the valleys of the Brooks Range. The thaw ended about eleven thousand years ago.

This new slump that we confront is an uncanny spectacle of slow yet dynamic motion, as if geologic time has been sped up, with the Earth under change before our eyes. The main slurry canal that freights the melted hillside to the river is reminiscent of Hawaiian lava flows, cold instead of hot, but a human-made crisis rather than a natural eruption.

Chris sets his video camera on the tripod to interview Gary for a detailed explanation. Like the rest of us, he wears a wool cap and several layers to stay warm in wintery conditions,[1] which seems discordant given the pile of melted earth that surrounds us.

"The permafrost is thawing at an accelerated rate," Gary says, as he turns and karate-chops toward the ice as if to call attention to a blackboard in a lecture hall. "Thermokarsts are found throughout the Arctic, but they're found with greater frequency and broader distribution as a consequence of a warming climate." It's no accident, he adds, that the slump faces south, and sun exposure has sped up the thaw.

1　The remote and automated temperature gauge on the eastern edge of the Noatak National Preserve recorded a chilly 44.4 degrees Fahrenheit average for the month of July 2022. Several days dipped below 32 degrees. On the warmer coast, Kotzebue averaged 51.9 degrees, 3.4 degrees cooler than normal and the coolest July since 2010. A good example of the increased variability and weather extremes in times of climate change.

On the vertical and overhung sections of wall at the top, he points out the active layer of permafrost ground—replete with roots and dirt and decomposed plants—that thaws every summer. Then, below it, he indicates the colder, glossy, blackened wall of permafrost ice under active thaw. He hucks a rock at the bulletproof ice, which repels his missile without a dent. Old as the Pleistocene.

As he talks, melted-out rocks continue to clatter down across the wall of ice and into what—we can now see up close—looks like a wastewater treatment pond. Rocks tumble in with a muffled slurp. We can't help but stare, gobsmacked.

"It'll continue to grow the rest of the summer, then freeze solid this winter," Gary says, "only to thaw again next summer and continue again next year and get even larger."

He talks about how large landslide thermokarsts can be destructive. For instance, upriver of the Iñupiat village of Selawik, Gary cites a landslide slump that has dumped metric tons of sediment into the Selawik River and negatively impacted their subsistence fishery. "It's not a benign landscape," he says, "and there's consequences not only to the vegetation, but to the wildlife and to the people."

National Park Service scientists say that while these large landslide thaw slumps are dramatic to look at, they only affect a small percentage of the Arctic landscape. After all, it's a vast place. But along with a myriad of other changes, the permafrost thaw has accelerated what ecologists call an old-succession ecosystem into an early-succession one—this brings in the shrubs that, in turn, bring in the moose that we'd seen the day before.

The losers amid this ecosystem of change are the caribou. Across the Western Arctic Caribou Herd's migration path, woody shrubs that flourish on the thawed landscapes have replaced lichen, an essential browse for the caribou. On the southern edge of the heavily thermokarsted Western Arctic parklands, lichens have drastically declined (new wildfires in the region have also stimulated shrub growth and prevented lichen recovery).

A thermokarst, Gary reminds us, refers to any northern landscape feature, large or small, where the permafrost has thawed and caused the ground to slump. This includes collapsed pingos, shrunken palsas, melted ice wedges, thawed bogs, flooded polygon land, newly made ponds, and the expansion of and disappearance of lakes.

From where we stand atop the thaw slump, high above the giant U river oxbow, numerous lakes appear beneath dark clouds. Many of these lakes have been caused by thaws that collapse and expand shorelines along with the lakes, until they reach old channels or low ground that drain to the river. Then the lakes disappear.

Since the start of records in 1980 (when ANILCA created the new park-lands), a myriad of lakes and ponds began to vanish in an irreversible trend—with a direct loss of aquatic wildlife habitat. These lakes drained, as could be expected, in warm years (from 2005 to 2007, and again in 2018).

We continue to stare at the apocalyptic landscape. It looks as if a bomb exploded on the riverside. Rocks crack down off the ice and slurp into the muck. Ancient permafrost glop-water gurgles and sloshes in slow motion down the immaculately carved canal and turns the Noatak River ashen gray.

The Bearded Ones
July 23-26, 2022

The next afternoon, our fourteenth day out, a Cessna 206 splashes down and sends a flurry of waves across the still waters of Lake Kavachurak. When the plane pulls ashore and the pilot shuts down the prop, I ask him for news of the outside world, but he merely shrugs. When I press him, he replies that he has no internet or television in Bettles. *Just as well, I think, because I come north to unplug, too.*

The pilot refuses to talk. He's either sleep-deprived, late for his next pickup, hungover, or depressed—not dissimilar from the other Bettles bush pilots I've encountered over the years (yet unlike the personable bush pilots I've worked with elsewhere in Alaska). I thank the Bettles pilot for the long flight, and he gruffly tosses me our resupply stuff sacks with ten days of food to get us to Noatak Village. Reliability always trumps cheerfulness in a bush pilot, whom your life can depend upon.

Gary hands me several days of leftover food and his spincast rod and tackle. He even agrees to trade his clean Kühl pants for my contaminated Eddie Bauer hot pants. A true friend, by the time he reaches Fairbanks, his thighs, too, will burn (but unlike me, he can soon buy another pair of pants). I reluctantly accept his can of pepper spray but let him take my heavy shotgun back out.

PREVIOUS SPREAD: Midway through the Noatak National Preserve—a six-and-a-half-million-acre park—we passed out of the high plains and back into the mountains with forests of spruce and cottonwood. CHRIS KORBULIC

"Be safe," he says, as we share a quick bear hug. *I will miss his easy friendship*, I think. Chris and I divvy up the food.

The plane shrinks in the immense sky to a size and buzz not dissimilar to a mosquito. It disappears as we portage back across the buggy tussocks to the river.

Chris prepares to push downstream, and before he leaves, I ask, for safety's sake, if we can keep each other in sight for the next couple of weeks on the giant river. But no sooner than the words leave my mouth, I know I'll need to make a retraction—he needs space.

A mile downriver Chris pulls over to wait for me. He's spotted a musk ox, so we walk up an ancient glacial moraine through head-high willows into a pungent tang, like a ripe barnyard—the smell of a bull's urine. Once we find the musk ox out in the open, I kneel down fifty yards away to look less obtrusive. Wary of wild animals, Chris stays in the distance.

The animal's fur robe drapes to the ground and sways in the wind, while he or she repeatedly rubs its hairy anvil-shaped head against its legs. Known as *uminmak* (the bearded one) by Iñupiat, the misnamed musk ox belongs to the goat family.

Flattened against its forehead, the horns circle back and bob out in sharp tips that resemble a sixties hot-curled hairdo. Through the telephoto lens of my camera, its canted eyes have a wise expression. Amid all the wildlife in the Arctic, the musk ox—like a turtle amid a bevy of rabbits— seems strangely misplaced. While the other large Arctic mammals roam and run hundreds or even thousands of miles, musk oxen perambulate fewer than a hundred miles a year. The animal in front of us—about seven hundred pounds—is in no rush, as if engaged in a stationary, dreamy contemplation of the world we share. I would do well to imitate it.

If sheared, I suspect the animal would lose a third of its bulk. The long, outer guard fur—underlain by wool (qiviut) softer than cashmere—is warmer and softer than my wool blanket and thick comforter at home.

In cold weather, their metabolism slows down. Their short, stubby legs, insulated by thickly draped robes, need little blood circulation.

To stay cool, every summer, musk oxen shed their qiviut underfur. But since this musk ox doesn't have the usual summer scraggly look, I can't help but wonder if the unusually cold summer has prevented the shed.

By now, any other large Arctic mammal would have sprinted away or charged us. But this musk ox is content to slowly mow the grass and linger here all day in the flats next to the river.

To watch a musk ox in a raw wind is to be transported in time. Although they live for only a dozen years, you can imagine their larger forebears along with the ground sloths and mastodons when they crossed from Asia tens of thousands of years ago. In all likelihood they first saw humans from the end days of Beringia, when fur-clad hunters stalked their larger ancestors with spears tipped with tiny, yet deadly, points chipped from obsidian and chert.

Suddenly, a second musk ox appears between the first musk ox and me and lets loose a long roar. I hear it as *"I'm alive. I am a survivor!"* It could also be a message to me as an intruder, so I don't approach any closer. It's a bull, likely in rut, which explains the barnyard stench. The other musk ox must be a cow—both sexes in the goat family have horns.

These musk oxen are generations removed from a herd reintroduced from Greenland after the whalers had slaughtered them—much like the hunters eradicated bison on the Great Plains. The extirpation, however, is not new. As the continental glaciers receded, climate change (and to a lesser extent, prehistoric human hunters) wiped out the bigger, helmeted musk ox, and scores of other megafauna species.[1]

Since musk oxen are found throughout the circumpolar Arctic, one could be tempted to call the species resilient. Yet, like the other long-departed Pleistocene megafauna, the greatest threat to musk oxen is climate change—specifically, the recent rain-on-snow events that ice

1   The old "overkill" hypothesis posits that the Paleo-Indian people who arrived more than eleven thousand years ago caused the disappearance of North America's megafauna—such as the giant beaver, the camel, and the flat-headed peccary—by overhunting. But since only mastodon and mammoth bones have been found marked with ancient, tiny bladed human stone tools, the latest hypothesis is that most of the megafauna were killed before humans had arrived by a change in the climate, such as the 1,300-year-long freeze of the Younger Dryas period 12,700 years ago.

With its magnificent robe swaying in the wind, a musk ox—placid and undeterred by our presence—grazes amid the dwarf fireweed and willows on a riverbank, Noatak National Preserve. JON WATERMAN

over the grasses. In past cold conditions without rain on snow, musk oxen could easily paw or nuzzle their heads through soft, dry snow to reach the salad bowl below, full of essential caloric and protein-rich energy to survive the winter. But as the Arctic continues to warm, the fate of the musk ox hangs in the balance.[2]

With eyes a-flash, the bull gives me another long roar that resembles thunder. I'm warned. Now he stares, too, so I look away and back off. Time for a quick bow and to give them some space.

———————

Within a few miles, we pull over, mesmerized yet again by a small thermokarst. On a high bank to the north, the river and the thaw have pulled away the earth to reveal a vertical wall of permafrost. The thirty-foot hill of dirty, rock-hard ice has been sculpted into whale-sized fins, dragon spines, and concavities that disgorge stones and clumps of earth. Just the kind of place you'd expect to find a mammoth tusk. We linger in the refrigerated air and take photos and videos.

As we paddle out of the thermokarst eddy upstream into the current, it catches the bows of our rafts and slingshots us downstream like fighter jets off a carrier. It pays to be playful in the sluggish packrafts.

The moody river has dropped again, despite last night's rain. Its surface has reverted to grayed cedar. Now, with little to do other than rotate our paddles through the air and back into the water, ad infinitum, we feast our eyes on the enormous press of land as the Noatak River Valley expands yet again before our eyes. Although we're only halfway down the long river, the mountains fall to the rear as we enter a broad plain.

———————

We leave the next morning under cumuli stacked like white chips on a blue poker table. The river continues its swirl and upheaval of pale silt. At a confluence, a dozen Arctic terns screech and chase and harry two loons that fly too close to their nests.

2   Grizzlies and rain-on-snow events aren't the only threats to the musk ox. In February 2011, a herd of fifty-five musk oxen were drowned in an ice tsunami in the Bering Land Bridge Preserve (due west of the Noatak River Delta). These once-rare events—called *ivu* by the Iñupiat—are increasingly frequent in the Chukchi Sea as abnormally warm storms sweep up from the Bering Sea. Then in fall of 2013, a rain-on-snow event on Banks Island, Canada, blocked the huge musk ox population from their iced-over grass. Twenty thousand musk oxen starved to death. Rain-on-snow events also cause caribou, moose, and Dall sheep to starve.

A golden eagle flies in to check us out; disinterested, it cocks its massive wingspan into a long swerve back downriver. The tawny bird flaps, then glides, and holds its wings up in a slight V, all its movements performed with a slow and powerful grace as it hunts for more likely prey: caribou calves, Dall sheep lambs, rabbits, ptarmigan, even a lemming would suffice. Within a minute, it's gone. In the Lower 48, golden eagles have a range of several hundred miles, but I can only believe that Arctic goldens might have to fly twice that distance to find food.

A lone caribou calf along the bank sprints away as it realizes Chris's paddle is not its mother's rack. It makes no sense to see a caribou calf this far south in midsummer, when most of the herd is hundreds of miles north. But everything is askew with the Western Arctic Herd, now two months later in its migration than it was on my first visit thirty-nine years ago.

Chris takes pictures and shoots videos. I sing to overlook my exhaustion. We pass a field of aufeis that cools the air and reminds us of the constant presence of winter, despite the thaw.

After an all-day push, we settle for a lumpy tussock camp. Since I'm spent, I flop out of the packraft onto the cobbles, too tired to stand. I stare at my heavy boat beneath the steep riverbank that separates us from the tent site and remind myself that I don't need another back surgery. Silent-yet-observant Chris, like a weightlifter at the gym, muscles my boat up the vertical embankment with little heed for his own back. No doubt he's tired, too, but as an experienced expedition partner, his thoughtfulness rules the day.

We deflate our boats and erect the tent near yet another set of gnawed antlers. It rains all night.

The dawn sky is tinged with shadowy menace. As we leave, dressed for full combat with the weather, a squall sweeps over us. Raindrops bead on the river and float across the water like opalescent pearls. I take

Exhausted from strenuous weeks on the go, often battling a cold wind, I fall asleep next
to my deflated packraft on our lumpy, tussocked riverside camp—while the indefatigable
Chris wanders with his camera. CHRIS KORBULIC

refuge under my rain-jacket hood, detached from worries about the weather, and happily watch the world float past.

Peregrine falcons hurl themselves out of their riverside cliff nests and swoop down at us as if shot from a bow: beaks become arrowheads, slate-blue wings the feathered fletching. They pull up short from their feints, climb back up into the sky, duck, then tuck back their wings, and shoot down at us again. All the while they shriek in their predatory, high-pitched *keeek, keeek, keeek, keeek!* Between the current and their high-speed dives, it all happens too fast to photograph. From a distance I look back and see their shapes flicker synchronously about the bluffs like stingrays that flap above the ocean bottom.

At the Cutler River confluence we stop at a half-hidden and dilapidated hunters' shack, its plywood boards festooned with nails to keep grizzlies out. A deflated and mangled white-water raft is jettisoned out back. Inside there's a 2019 calendar, a book on keto diets alongside a worn-out mattress, and stacked willow firewood with a flue pipe to nowhere that renders the cabin forlorn without a woodstove.

Back in the boats, the Cutler River influx holds a clearly delineated blue entrance lane into the dark highway of the Noatak. Within a half mile, the dark silt traffic of the Noatak rush hour envelops the Cutler.

We're on a long southern loop of the river, a couple dozen miles north of the border of Kobuk Valley National Park and its extraordinary sand dunes. That park could easily absorb the state of Delaware and abuts yet another protected and (even larger) wilderness landscape, the Selawik National Wildlife Refuge. There, on the northern edge of the boreal forest, amid the southern reach of the Brooks Range Ice Age, the glaciers carved out some twenty-one thousand lakes. You can get dizzy with the scale of the landscape here. Wildlands stacked upon wildlands—thanks to ANILCA.

This day I'm overjoyed to find my strength and set the pace. After two ibuprofen tablets, I even feel rejuvenated—propelled by hard-earned mastery and muscle memory—despite my aged body and its sixty-six revolutions around the sun.

The riverbanks scroll by us as fish dimple the eddies. Ospreys screech. As the landscape widens, the river expands to hold it all.

In late afternoon I cast into the clear water of Mapik Creek where it mixes with the Noatak, and within minutes, I have a scrappy grayling on the line. It spins, flips, and splashes out of the water, a born fighter. Nearly eliminated from Lower 48 waters through habitat loss, I will always be grateful for the grayling—its shiny armor scales, its pink, pale blue-spotted sail fin, its small pouty mouth. Grateful, we eat crouched in the rain away from the tent, where the wind is broken by a copse of cottonwoods.

It's always a surprise to see cottonwood trees so many miles north of tree line. Researchers discovered that cottonwood trees moved up into the Brooks Range sometime before eleven thousand years ago as the Ice Age ended. Eight thousand years ago, for some unknown reason, the trees again disappeared from the region, but in the recent past, as the climate warmed, they returned. I'm glad, too, as we zip up the tent, soothed by a wind that comes through the leaves like the patter of rain. *Without trees*, I think as I shut my journal, *we wouldn't hear the wind so well*.

We paddle hard all the next day pushed by a bitter north wind that sweeps down the river. Every two hours we stop in a willow windbreak, stretch for fifteen minutes, and re-energize with the day's spartan ration of nuts and energy bars. On our late-afternoon stop, we jog in place to stay warm as we finish our coffee. Although my thermometer reads in the mid-forties, we subtract another 20 degrees for the windchill. Giddy that we're able to defy the cold and that the wind has become our ally, we're partially insulated from the cold by our spray skirts. We push our bulbous boats back into the wind tunnel and agree to continue for at least another two hours.

"But if another cabin appears," says Chris, as he shoves backward into the current, "let's take it!"

Ten minutes later, the long shot comes true: a red-walled tin-roofed cabin appears a half mile north. Without a word, we exchange a quick glance, a nod, and immediately check it out.

Grizzly fur sprouts like errant whiskers from the cabin corners where bears have scratched their backs. The door is unlocked and, honoring the proviso throughout Alaska that cabin "latchstrings are always out" to welcome wilderness travelers, we unscrew the bear board shutters that bristle with nails. Inside we find a propane heater—so we crank it up and hang our wet clothes.

We then discover the motherlode of long-expired food in the kitchen. After weeks of exertion in the cold, on freeze-dried rations, it feels like we've hit the jackpot: granola bars, soups, Dinty Moore beef stew, Oreos, Ritz crackers, spaghetti, pasta, Fig Newtons, ancient ground coffee, beans, gummy candy gone to rocks.

No matter that we could break a molar on the candy, that the softened crackers would suffice for toothless codgers, or that the canned food might impart botulism. We pull out each familiar brand-name box or can from impossibly overfilled pantry shelves and declaim the expiration dates:

"Rice Krispies!" I say, "1999."

"Hunts Ketchup," Chris retorts, "1994!"

"Milkman," I say, "1991!"

"Look," Chris says, with the grand prize, "mac and cheese, almost as old as me: 1988!"

In a giant fry pan I sauté an ancient can of Spam, then boil up and mix in three boxes of calcified mac and cheese—I season it with a smorgasbord of caked spices that I break into powder with a knife. The meal

Our discovery of vintage Spam along with well-aged mac and cheese—along with having a cabin roof over our heads amid the icy wind and rain—proved a welcome, albeit crunchy, caloric recharge. CHRIS KORBULIC

would feed half a dozen mouths under normal circumstances. We easily finish it off.

We stay up late to read the journals of the Kotzebue family who built the cabin a half century ago. Too remote and inaccessible for most Iñupiat, the owner, Warren Thompson, barged up the lumber and other supplies to build his house and two sheds. Thereafter, he used a snow machine or flew his family up from their Kotzebue home in his small, fat-tired plane that he landed on an adjacent grass strip. Like all rural residents,[3] Warren, a Caucasian (married to May, an Iñupiat), had subsistence rights that grandfathered in his cabin after ANILCA took effect. His forty acres also lay within wilderness-legislated preserve. The journals depict epic hunts and aggressive grizzlies and children who swam in the meander of a (now dried-up) river channel as their ancestors had done over the centuries.

With the deaths of Warren and May more than a decade ago, the cabin sees scant use. But the beds are neatly made and the floor swept as if the owners left yesterday. We're grateful for the shelter and leave it cleaner than we found it.

3 One of many concessions awarded to Alaska in the ANILCA legislation was that all rural residents—rather than just Indigenous Alaskans—who live close to the newly protected conservation lands had subsistence user privileges. Unlike in the Lower 48, subsistence users can deploy snow machines, airplanes, or motorboats in some Alaskan wilderness areas.

Salmon
July 27, 2022

1 I was told by a psychic—through an unsolicited correspondence—that this bird is my spirit animal. The letter writer had no idea (unless he really did have psychic powers) that an early Audubon print—my only Audubon—of the gyrfalcon hangs on a prominent wall where I see it every day that I'm home.

Although a second night in the cabin is a sore temptation, the lack of rain or snow sends us back to the river. A relentless headwind makes us avert our faces and paddle doggedly.

As we pass through a river gorge, a pair of ghostly gyrfalcons glide silently off a cliff ledge and pendulum and stoop and mirror one another in flight. Unlike the raucous peregrines, the gyrfalcons have no interest in unnecessary conflict and keep their distance from us. I'm awed[1]—they're the largest falcons in the world and the only bird of prey that stays north for the winter.

As the wind reverses direction, we all but fly downstream to the Nimiuktuk River. It's fringed with the greenery of small trees along the water. Up above, the pale green tundra is abruptly circled by the rocky, tonsured mountainsides—nature's version of male pattern baldness (or so it occurs to me as I scratch my hairless head). On a flat gravel riverbank alongside the trees is a pale red and glossy green tent, conspicuous amid the natural foliage. Since these are the only other boaters we've seen in days, and since the wind has worn us down, we pull over to chat.

Thomas Hassler and Michael Grohmann—Italian and Austrian fish ecologists, respectively—have flown halfway around the world to get to the Noatak. Since European rivers are all altered and dammed, Thomas

PREVIOUS SPREAD: On a blustery Arctic day, lost to the wonder of moving water and surfacing fish, Michael Grohmann sends his fly out into the confluence of the Nimiuktuk and Noatak River waters. JON WATERMAN and Michael were drawn here to experience a river untouched for its entire length. They have also come to fly fish and to dine on salmon eradicated by more than a million dams in Europe—where the fish is considered a delicacy. You can detect a nobility about their mission, as if this is a last hurrah for knights-errant dedicated to fish and rivers in a world more focused on development than preservation.

On their smoky, twiggy fire they brew us cowboy coffee and we sip it gratefully from their enamel cups beneath angry clouds that race through the ambivalent sky. Our noses run in the cold, and we stomp our feet to stay warm. We swap tales of the wildlife we've seen and banter with the evangelical rapport shared by like-minded wilderness pilgrims. They tell us that the silt added to the watershed—caused by all the thawed permafrost features they've seen—is likely to damage the fish.

But the Nimiuktuk River, for now, remains pellucid as thick glass. More than forty miles long, the Nimiuktuk (Iñupiaq for cottonwood trees) drains the De Long Mountains and hits the Noatak between the mountainous Aglungak Hills to the east and the bald Kingasivik Mountains on the west. The cold, limpid river is one of several Noatak destinations for two Europeans and thousands of salmon. In one of the universe's most wondrous migrations, the fins that briefly stick out of and riffle the river alongside the twiggy fire belong to salmon that use the Earth's magnetic field like a compass out in the ocean and then employ a smell memory to return to their birth river. In the eddies alongside us, females pair with males and swim a couple dozen miles upstream to find a gravel bed swept by shallow water. Often in the same square footage of its birth, the female will scoop out a redd (spawning bed) with her fins and deposit several thousand eggs that are immediately fertilized by the male's sperm. The female then sweeps gravel over the eggs to hold the eggs down and to allow fresh water to pass through. Their life's mission complete, the spent salmon die and litter the shallows. Cue grizzlies and eagles.

Alongside us, tracks of a grizzly sow and cub pock the silt banks. They're maybe several days old. But as a group of four we feel a certain safety in numbers, and Michael and Thomas are happy to have us camp with them.

More than a hundred cliff swallows have commandeered an old riverbank, and they pop in and out of the shadowy depths of their cup-sized nest holes. They pluck mosquitoes from the air around our tent as we tie on lures and flies and attempt to pluck fish from the stream.

This far north, the salmon are mostly chum (*iqalugruaq* in Iñupiaq) that spend up to four years in the Bering Sea. As the second biggest of the five Alaskan salmon species, the chum is the most widely distributed but the least desirable, given its bony flesh. It's prized among the Iñupiat, who net chum out in the ocean beside the Noatak River Delta, when the silvery fish are still firmly muscled and palatable before the long swim upriver. Elsewhere in Alaska—where I routinely fished for king, sockeye, pink, or silver salmon—anglers call chum "dogs" because they're fed to the huskies. I don't mention the slur to our European friends.

Michael and Thomas wander upriver dressed for winter; balaclavas hide their faces. They cast with precision and study each eddy as they lose themselves in a world of water. Utterly focused on their mission.

A couple of hours later they return with two ten-pound chums. The chum have lost their metallic-blue ocean color and, unlike other salmon, have turned green with distinctive pink calico splotches down their flanks. Their upper jaws are hooked into kypes with needlelike canines.

Since I've been thwarted in the Nimiuktuk with the clunky spincast rod, I move downriver and cast into its colloidal confluence with the silted Noatak River. On each of my seven casts, I catch a grayling to supplement our salmon dinner. After a brief study of their divine, golden pupils, I carefully remove the hook and return all but the two biggest fish to the river. The keepers barely weigh a couple of pounds.

I reach for more fish eggs, while our European friends Thomas and Michael crowd the fire
for warmth and contemplate their next move amid rice, fish-head soup, and grilled salmon
and grayling. CHRIS KORBULIC

Thomas cuts off the salmon heads for fish soup and boils them in our fry pan. Michael removes the skeins of eggs for caviar, and we grill the grayling and salmon over the fire. The smaller grayling cooks quickly. Famished, I find that the firm white flesh flakes apart beautifully with my spoon and tastes faintly of thyme—like trout, without the nutty flavor.

We spatula off hunks of meat from the thick steaks of chum on the grill. The rosy-gray flesh is soft and tastes bland, but not oily like other salmon. The unavoidable tiny bones that come loose in my mouth could double as small toothpicks.

I try a spoonful of the fish soup, which immediately prompts me to guzzle some cold river water. Michael wears a wool hat pulled down to his eyes and a Cheshire cat smile. "I put lots of chili powder in," he says, "which should help keep us all warm."

It happens that the chum are renowned for their red caviar, so distinctively red they look like they could glow in the dark. In fine sushi houses this red caviar is marinated in soy sauce and sake and called *ikura*. Pressed against the roof of your mouth, the raw eggs pop open and spread a unique and creamy texture and taste over the tongue. The Japanese call this savory fifth taste umami—neither sweet, sour, salty, nor bitter. Although deprived of vegetables and other nutrient-rich foods on our trip, the eggs, more than any other part of the salmon, are rich with both protein and omega-3 oil.

In lieu of a reservation at my hometown's pricey sushi house (where my sushi-chef son cuts me a discount), I'll take the free red caviar out here. We stink of campfire smoke and fish guts, our hands are blackened by soot, and we're 260 miles from the nearest major road, the dirt Dalton Highway, but we have dined like royalty. I pass around dessert—decades-old chocolate bonbons—from the cabin. The expired food story makes our polite European friends laugh.

"This is the life, right?" I raise my cup of tea and our metal cups clink like the swords of four knights-errant.

Despite the bonhomie, I know that wild Alaska salmon is in jeopardy. After Indigenous Alaskans had repeatedly told fishery officials that both the catches and salmon sizes had shrunk, scientists conducted a study.[2]

Based on measurements of 12.5 million Alaska salmon from 1957 to 2018, the study concluded that fish spend less time at sea and return earlier to spawn, which results in smaller-size salmon. Chum have shrunk by nearly three inches (or more than 2 percent), while the commercial catch has also been greatly reduced.

In the study, the scientists noted that climate change has contributed to smaller sizes in cold-blooded water creatures across Europe, in sheep in Scotland, and in migratory North American birds. It's also believed that that the competition between the wild and hatchery salmon throughout Alaska may have affected the fish.

Elsewhere in the world, wild salmon populations have diminished through climate change, habitat loss, dams, and development. Until now, Alaska has remained a refuge for the salmon that have swum the oceans for ten thousand years.[3] Their incredible voyage as two-inch fry out into the Pacific, and back as huge salmon to their precise birth stream, provides essential nutrients to the ecosystem, bolsters the Alaskan economy, and supports the age-old culture of Iñupiat and other Indigenous people throughout the state.

Until now.

2 King (Chinook) salmon sizes have declined an average of about four inches, with a loss of 8 percent of their body size. Red (sockeye) have lost about one and a half inches, a 2 percent decline. The most rapid size changes occurred over the last decade. I learned this before our Noatak trip from research for my National Geographic book, *Atlas of Wild America*.

3 The million or so chum that spawn in the Noatak and Kobuk Rivers hardly compare to the big Bristol Bay fishery that chases hundreds of millions of salmon. "Our Kotzebue Sound fishermen used small outboard-powered boats and fished by themselves or with their sons or brothers," wrote Iñupiat Willie Hensley of ANCSA fame, whose boyhood summers revolved around salmon. "A $10,000 summer would be a big year for most fishermen, but occasionally they earned $20,000 to $45,000 in high years."

But for the years 2018 to 2021, after the National Oceanic and Atmospheric Administration confirmed that climate change had disrupted Alaska's salmon runs, the US secretary of Commerce declared multiple salmon fishery disasters across the state and released $66 million in federal reparations for economic damages.

Nautaaq

July 28–August 1, 2022

Shortly after we leave the river of cottonwoods, a herring gull repeatedly swoops above what appears to be a blond boulder in the distance. As we paddle closer, the boulder moves and becomes a grizzly. A strong river current pulls us inexorably down the single channel that piles up against the bank directly beneath it. The farsighted bear gets a bead on us right away—despite an upriver wind—and immediately vaults down the steep embankment in a cascade of rocks and dirt onto the shore. Where it stands in wait.

I pull out the camera, confident that I can out-paddle a grizzly if it jumps into the big river's surge. The subadult resembles a jowled, woeful-eyed Saint Bernard, fuzzy ears up attentively, with blond head and chest fur elegantly contrasted with dusky brown fur.

One "Hello, Mr. Bear!" changes the whole dynamic: the grizzly snorts and promptly explodes up the bank and out of sight into the low willows. No doubt bound for easier spoils at the nearest salmon stream.

We shoot past tundra banks undercut by the river into dark tunnels. (Villagers tell me later that they never saw undercut banks before the new millennium.) Paired-off loons fly upriver and laugh down at us in crazy tremolo. Fireweed blankets the hillsides in vivid merlot.

PREVIOUS SPREAD: Noatak village and the ancestral lands of the Iñupiat amid wildfire smoke. The river washed out a road near the airstrip, used for vital food resupply, since barges can no longer transit the flooding river. JON WATERMAN / LIGHTHAWK

This day we manage to crank out several dozen miles with our paddles. By evening at the arboreal Kaluktavik River (Iñupiaq for the place where wood is available), I'm too tired to fish, and lie in the sleeping bag and drift off into dreamland with a smile amid a chorus of peregrines, sparrows, gulls, and warblers.

I dream that I'm imbued with avian powers and—arms held wide—I lose myself in rapturous flight that allows me to explore the Earth effortlessly. It's my favorite, most recurrent dream. If I could dream it every night, I would foreswear alcohol and dessert, and maybe even trips to the Arctic.

In a morning christened with abundant, lurid sunlight, we slip past patches of forest that exude warmth. We hear the susurration of water in motion against its banks and over the rocks, and the motion creates a wind that passes through the spruce trees to give the river further voice: steady and calm, the sound of energy and full-blown joy.

Off in the distance, Chris sees five wolves—four pups and the mother—lope into the brush. A musk ox, the tawny shepherd of Beringia, climbs slowly up a steep riverbank and disappears before I can pull out the camera.

We stop beneath a giant thaw slump. From park service records and our overflight, we know that this thermokarst first began to slide into the river a dozen years ago. But the landslide has dried up and the exposed gravel has begun to grow over with willow, alder, dwarf birch, and cottonwood. Tiny lemmings scurry about our feet. From on top, several hundred feet above the river, the partially revegetated dirt slopes resemble a clotted scab about to heal over with new skin on the Earth's surface. The forty-foot headwall is too steep to hold vegetation and looks like the head of a gravel pit scraped clean by a steam shovel.

Down its middle, two perfectly graded and immaculately scooped-out canals show where the once-active thermokarst sluiced melted permafrost sludge down to the river. Neat as conveyor belts. Unlike the last thaw slump, I leave with an impression that, given a chance, the Earth could endure, even though the Arctic thaw is already well underway.

Back in our boats, the river stretches a mile wide with innumerable, braided channel choices and meanders. We take turns in the lead and trust each other implicitly with problematic turn choices at channel intersections. Since the river is filled with silt, it's difficult to read water depths, but by virtue of the enormous water volume, we seldom ground out.

We paddle through forests or around spruce trees where the Noatak claws at the banks and permafrost subsidence has collapsed the ground and widened the river. Where the current speeds up on the outside of the curves, the river scours the banks and sends sediment downstream to accumulate on the inside of the curves. Gradually, over the years—dependent upon the strength and repetition of flood events—the inside of bends fill with sediment that builds new lobes of land. From a bird's-eye time-lapse view, the river coils to the ocean like a snake in constant conspiracy with the land. Over the decades, the sun-warmed gravel on the inside bends prompted the growth of brush, then cottonwoods, and eventually, spruce trees. It's hard to know how much this natural riverscape succession is accelerated by permafrost thaw and climate-change-generated floods.

As an osprey alights on its nest, I shout to Chris and tie my packraft off to a tree. Long-lensed camera in hand, I push into the forest, thick with spongy sphagnum moss and fragrant Labrador tea and air that bristles with mosquitoes. A hundred yards in, up in the nest, three chicks play dead as the mother glares down at the obnoxious photographer.

Three alert ravens swoop in to assess whether my intrusion will allow them to get to the chicks, but the osprey immediately takes to the air and drives them off with high-pitched whistles answered by the ravens' hoarse pandemonium of squawks.

Time for me to leave. I forgot my bug dope and bear spray.

We know from the map, and from several motorboats headed upriver, that we'll soon reach Noatak Village (Nautaaq in Iñupiaq). Although we wave to the boats as they speed by, no one waves back. It might be that our custom of a hand wave is not the Iñupiat custom. Although we would enjoy a chat if they stopped, it occurs to me that they might choose to leave us alone out of respect for our wilderness experience, which is very different from their experience in what they perceive as their backyard.

At the entrance to Noatak Canyon, we hang tight along the inside bend to avoid a thunderous conundrum of rapids. The limestone walls pinch the river into a flow that feels like the Colorado River—minus its unavoidable rapids—where it pulses through its own Grand Canyon.

We stop to match an old photograph with the modern-day view and find, as expected, that the former tundra slopes are now overgrown with a dense grove of high alders. Sometime over the last half century, a ginormous section of cliff has collapsed into the river.

We float past a beach of sand dunes with a picturesque and unlikely stand of spruce watered by a brook—as if we've just been transported to the moist Pacific Northwest. Several miles downriver we get out again to investigate an Iñupiat homesite alongside a large stream. For orientation, we climb a ladder up a high, precarious tower with thirty-foot-tall log pillars plunged into unstable permafrost. From on top of the platform above the spruce forest, we get an exceptional view of the landscape. On the edge of the tall boreal forest, small trees have taken root and, eyes squinted, it appears as though the trees are in the process of a slow migration north—like fireweed blossoms that unfold and open on their way up the stalk as the season progresses. To the east the Maiyumerak Mountains loom several thousand feet above the forested plain with the steely, surreal magnificence of a Bierstadt Rocky Mountain oil landscape.

The homestead is an impressive tribute to its owner (Ricky Ashby, according to boxes in the vestibule). Without a crane, it's hard to

Chris and I were routinely invited into Iñupiat homes, fed, and treated like old friends. Or royally beaten in games of Rummikub. We would never forget the generosity of the kind people of the North. JON WATERMAN

imagine how he raised the tower's four poles. Ashby's cabin is neat as a pin. Unlike other preserve hutches, he has dispensed with nail-armored shutters—we later learn that he's never had a problem with grizzlies because he keeps his meat and fish frozen in a permafrost cellar. The unlocked doors open with a clever latch system that bears have not yet mastered. Tools hang gracefully in a shed alongside a handcrafted canoe. We walk across the property on giant, flat stone steps, carefully selected and hauled from some distant site. Another small cabin has a woodstove and serves as a sauna. And the outhouse doesn't stink.

This is Ricky Ashby's sanctuary—unlike hunter shacks elsewhere in the preserve. Rather than intrude, we pitch our tent on an island across the river.

As I lie awake on my back and consider the last three weeks on the go, I realize that I haven't slept as well in years. It's the lack of alcohol combined with the strenuous days, as well as the abstinence from a computer and the internet. Mostly, though, it's the peacefulness of the northern wilderness and the tranquil burble of the river that lulls me to sleep.

On the way to Noatak, a pair of sandhill cranes fly so low and close that we can hear the wind vibrate through their feathers. The long-lived species first startled me as they flew above the headwaters half a lifetime ago. I have gone to lengths to watch the majestic sandhill cranes. As the two cranes moan and honk into the distance, I wonder if I'd seen the same pair before, amid the thousands, in their winter migratory grounds at wildlife refuges in New Mexico, at Bosque del Apache along the Rio Grande, or in southern Colorado, amid the barley and wheat fields of the San Luis Valley. Stranger things have happened.

Pink polka-dotted Dolly Varden repeatedly rise around us and send out ripples in concentric circles. We're startled by a beaver that slaps its

1 In a May 2022 *Scientific Reports* article about the period from 1949 to 2019, researchers assessed 1950s aerial photos and found no beaver activity in the Western Alaskan Arctic. Then 1980s satellite imagery showed the first incursion of beavers. By 2018 (thanks in part to the new woody shrubs that provide food and dam material for the beavers) the ponds had doubled; more than eleven thousand new ponds appeared across the tundra.

Ricky Ashby routinely drinks from the stream alongside his cabin and recently contracted giardiasis from the beavers.

tail with the percussive splash of a boulder dropped into the water. For the next few miles to the village, we see scores of gnawed punji-stick tree saplings on river right.

On drainages alongside the river, the beavers' dammed-up ponds act as a heat source that further thaws the permafrost. Twenty years ago, there were no beavers in Arctic Alaska. Now they're ubiquitous in the lower river lands, attracted by the new warmth, and as researchers attest, they accelerate the thaw with ponds that absorb and hold heat.[1]

We pull into the large eddy that serves as an impromptu marina for Noatak. It's obvious that the half dozen boats we've seen upriver are only a fraction of the local fleet, parked for months due to the exorbitant price of fuel. For a village of nearly six hundred who depend on boat travel—to gather firewood, hunt, fish, or pick berries—this would be the disastrous, olden-day equivalent of scuttled kayaks and dogless sleds.

At first, we walk as uneasy strangers through the village, until Iñupiat stop and shut down their four-wheelers to say hello and ask where we've come from. In the Native Store, people repeatedly smile and say, "Welcome to Noatak!" We feel appreciated and respected.

We're directed to Hilda and Thurston Booth's home, where we're greeted like old friends. Over reruns of *Gunsmoke* that blare so loudly on the wall TV that we have to read lips and raise our voices to be heard, Hilda says we can stay at her brother-in-law Ricky Ashby's home, since the Booths already have a house guest.

We haul our boats to Ricky's house and are relieved to find that our host—as befits a bachelor content to be alone in his distant wilderness cabin—does not own a television. And he has a wealth of stories that easily supplant a TV.

It takes him three days to trek the fifty miles alone to his cabin when the river freezes over. In recent years, the delayed winters of the climate crisis have forced him to wait until December instead of October.

The well-traveled Ricky Ashby at home in his hand-sewn caribou fur and wolverine-ruffed parka and mukluks that Iñupiat have fashioned over the millennia. CHRIS KORBULIC

After a day of gathering the Iñupiat version of spinach—sourdock (*quaġaq*)—on the outskirts of Noatak Village, Ricky chops it up with his ulu and then will boil it to remove the sour-tasting oxalic acid. CHRIS KORBULIC

As a devout Quaker, Ricky has traveled to South America, Europe, and many Lower 48 cities to participate in community prayer sessions known in his chosen faith as "gospel trips." Since he doesn't have a boat, he's stuck in the village, where he makes bone-handled knives and sews traditional fur jackets and mukluks. Nor does he own a four-wheeler, so Ricky takes epic walks to look for berries or to gather bushels of the coveted sour dock. Iñupiat eat the large-leaved plant like spinach, chopped up and boiled with sugar.

Compared to his comfortable cabin sanctuary surrounded by the wilderness of the preserve, his plywood-floored, one-bedroom Noatak house is a temporary crash pad. Sad-eyed, yet chatty, Ricky prefers a solitary life that, he explains, stems from the abuse he suffered as a child. His village-time salvation, as it is for most Iñupiat, is found through family and the Noatak Friends Church. When he mentions that there are still shamans around, he says it buoyantly, without Christian condemnation, as if perhaps his people still embrace the old ways.

Ricky has two bookcases filled with volumes about Eskimos and adventure in the North. He pulls out his own handwritten English-Iñupiaq dictionary that he updates with forgotten words. Unlike most homes in the village—filled with more contemporary bric-a-brac and appliances and heated with diesel oil[2]—Ricky heats with a woodstove. His artwork, proudly displayed on the walls, serves as a window to an almost-vanished way of life in the North. He explains the tool artifacts on the wall and proudly refers to himself as an Eskimo—just like the people who originally made the tools.

A two-by-two-foot collage has a background canvas made of seamed blue jeans, framed with dark, hand-carved wood. The collage displays immaculately made and symmetrically curved ulus, harpoons, arrowheads, and spearpoints with ivory or bone handles that could have been gathered from an ancient hunter's caribou tool bag.

[2] The town's electricity comes from the power plant, which runs entirely on flown-in diesel fuel. The cost of electricity is nearly a dollar per kilowatt-hour, the most expensive energy in the entire region, and nine times what I pay in Western Colorado for electricity. By the end of 2023, the federal Office of Indian Energy Policy and Programs will complete a $2 million solar-panel installation in Noatak, expected to reduce the village's diesel fuel consumption by 10 percent.

A larger piece of patchwork art above the kitchen table shows bow drills and bone fish lures and other forgotten Iñupiat implements that Ricky has carved and whittled himself. On the several-inch-wide aspen frames that surround the tools, he has inscribed a scene of the mountains that rise above the sparse forest, his cabin, and the tower next to the stream, along with careful drawings of two dozen other tools invented by his ancestors and their names—*kuvraq* (net), *qayaq* (kayak), *tagluk* (snowshoes), *tuuq* (ice chisel), *iktaq* (hook and line).

At the top of the framed piece of kitchen art, Ricky wrote in the God-driven meaning of his life, from Ecclesiastes 3:1: "To everything there is a season, and a time to every purpose under the Heaven." Underneath is the Iñupiaq translation: "*Piuiqut saupayaaq Ataani qilaum.*" In concert with the old tools, the verse shows the divergent world that the Iñupiat were marched into more than a century ago.

As per a skewed 1880 US Census report, four hundred Eskimos lived here in an unincorporated village, but when more white explorers passed through five years later, they saw only an abandoned camp. Noatak next appeared in the 1910 census with a more realistic count of 121 residents (who lived in the village first established two years earlier).

By then missionaries from the Religious Society of Friends, or Quakers,[3] had already begun to radically change the lives of the Iñupiat for miles around. In 1890, northern Alaskan Natives practiced their rich and traditional spiritual lives without Christianity, so Sheldon Jackson—a Presbyterian minister appointed Education Agent for Alaska—carved up the territory into regions designed to prevent competition between the various denominations that would begin Christian conversions. The Friends took northwest Alaska, home of the so-called Eskimos.

The conversion of the people the Bureau of Indian Affairs called barbarians had been perfectly timed. White whalers had decimated the bowhead whales and walrus that Iñupiat depended upon, while the

[3] The derogatory term *Quaker* came from the widely held view in seventeenth-century England that the Friends trembled and shook when they confronted God or the devil.

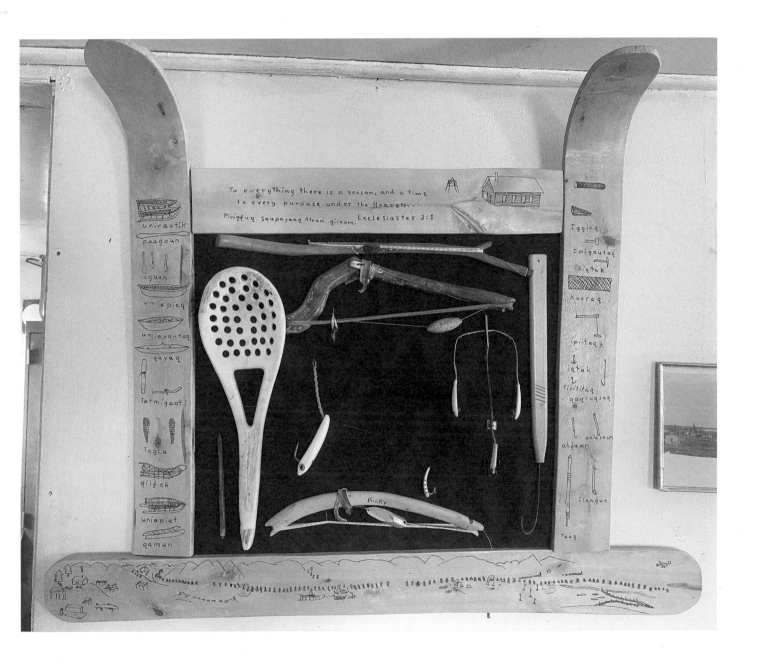

Ricky's art shows tools Iñupiat used out on the land. This assemblage is headlined with the famous verse from Ecclesiastes, and its Iñupiaq translation. On the horizontal frames he drew scenes from his wilderness cabin upriver. CHRIS KORBULIC

Western Arctic Caribou Herd had also declined. In addition, whalers had brought diseases that the people of the Arctic were not immune to, so starvation and death swept through the North. Greatly weakened, Iñupiat were primed for salvation.

In 1897, the Friends Church missionaries Robert and Carrie Samms, along with Anna Hunnicutt, arrived in the middle of an Iñupiat trade fair (attended by people who traveled from as far away as Siberia for the annual event). Sheldon Jackson had sent the missionaries to build a church and continue the groundwork of an earlier proselytizer. Since the church mandated that male missionaries had to be married, Robert (32) had asked for the hand of Carrie Rowe (19). They arrived just months ahead of a flood of gold miners. The missionaries' collective food and supplies attracted the more desperate Iñupiat who moved in from their camps up the rivers and along the coast. Several years later, Jackson sent two hundred reindeer and put Robert in charge of a Kotzebue herd that would soon grow and help alleviate hunger among the Iñupiat, whom Robert trained to become herders.

A mechanic with a pinched, narrow face, a handlebar mustache, and a severe gaze, Robert continued to spread the gospel inland and along the coast from their first church in Kotzebue. Combined with Presbyterian and Episcopalian ministries, Christianity would rapidly spread as far east as the Mackenzie Delta in Canada and systematically upend the ancient beliefs of the Iñupiat and Inuit.

At the time, the Friends were more dominant in America and dedicated to the "Inner Light." As pacifists, Quakers in the United States were opposed to slavery and actively supported equal rights for Blacks and the women's suffrage movement.

The charismatic Sammses visited the ritualistic, shaman-led drum dances along the coast and preached against the evils of alcohol, tobacco, polygamy, and other vices. Iñupiat shamans were revered healers and storytellers who flew outside their bodies into the afterworld and encouraged their followers to abide by the old ways. At one drum-dance event that typified the conversions, Robert asked why *akutuq* (Eskimo ice cream, a mixture of seal fat and blueberries) had been smeared on the ceiling. When an Iñupiat told the missionary that the traditional food had been left on the ceiling in the usual custom for one of several devils they believed in, Robert flew into a rage that ruined the dance.

The Sammses began to prohibit the drum dances and taught the Iñupiat biblical hymns, which they sang in English as Carrie played her harpsichord and Anna (who would marry a gold miner) played the violin. Their techniques proved incredibly effective. They demonstrated how modern medicine could heal people in a way that the shamans could not. As the Sammses learned some Iñupiaq (the Iñupiat language) and were given their own Iñupiaq names, they made their converts pledge to abstain from liquor and to follow a course toward the Inner Light. Converted villagers, in turn, were empowered as lay missionaries and spread the good word to other Iñupiat villages on gospel trips.

The missionaries routinely broke formerly sacred taboos. For example, they encouraged the locals to burn the platform graves (set up on two tripod poles, with the body tied on a top horizontal pole in a burlap sack with all their belongings below on the ground) where dead relatives were left outside the villages. When the Iñupiat saw no harm befall either the missionaries or converted Iñupiat, more people began to break the ancient taboos. Those who drank, smoked, or had sex out of wedlock were ostracized.

To complement what they'd done in Kotzebue, the Sammses built a mission school in 1908 and brought in a teacher to the place that would be named Noatak. Attracted by education for their children, the nomadic Iñupiat moved from camps up and down the river and built sod-log houses in what would become their new home. Although the Sammses followed the disciplines of their own peaceful faith, their actions also reflected the 1899 mission statement of the Bureau of Indian Affairs: "The Indian must be prepared for a new order of compulsory education, and the traditional society of Indian groups must be broken." The goal was to create individuals influenced by the doctrine of Christianity, rather than communities guided by their shamans.

To this point, the people of the North believed that their souls were indestructible and that they were repeatedly reincarnated into new humans. As animists who acknowledged many gods, Iñupiat also believed that the land and all animals had their own souls. Yet the Sammses eventually convinced the Iñupiat that there was only one God, no reincarnation, and the one ultimate destination, heaven. Rather than deny that multiple gods existed, the persuasive Sammses convinced their Eskimo charges that multiple gods were the spirits of the devil. And that the formerly revered shamans practiced devil worship.

The Sammses took it one step further as they began to teach the English language. To learn the Bible, to find the light within, the Eskimos had to learn English. Never mind the ancient language that Iñupiat had spoken for many centuries before Britons began to speak English in the fifth century.

Iñupiaq (or Inuktitut) is a language of precise nuances and environmental clarifications of the Arctic world. The Iñupiaq language had more to do with space than the English language's preoccupation with time. Northern hunters spoke slowly to make sure they were understood, without hand gestures or filler words like "aah" or "ummm." Like other cultures colonized throughout the world, assimilation and cultural degradation began as the Iñupiat lost their language.

Eventually, the Bureau of Indian Affairs[4] took over the business of education and renounced the Quakers' ideological motives. Still, assimilation continued apace. New missionaries and teachers arrived. As children, Ricky Ashby's grandparents and parents (and eventually, Ricky) were punished or beaten if they spoke Iñupiaq in school. This practice ineffably broke the gentle people. Prior to the missionaries, Iñupiat did not punish their children.

4   The Bureau of Indian Affairs' treatment of Iñupiat and other Alaskan Natives followed the history of Native American assimilation and dispossession in the Lower 48: educational failure, dependency on government welfare, poverty, health issues, and rampant suicide rates. Yet thanks to ANCSA, there are no reservations or treaties.

The Sammses took lengthy furloughs and moved to Kivalina on the coast. In 1940 the first post office arrived in Noatak Village. Although the Noatak Friends Church would remain and the people of Noatak were largely converted, like many Iñupiat in remote villages, they still retain a strong sense of cultural identity. The village would remain a place where Iñupiat could still connect with the land and its animals. With conversion partially accomplished, the people had been indoctrinated to embrace a new kind of capitalistic wealth, different from what Iñupiat originally knew as the riches of the Arctic (before mines and oil). But a cash-barter economy persists, and while traditions changed, they were not lost. Today Iñupiat Studies is a required course at the school, but few elders still speak Iñupiaq. Friends churches are now found in ten northwest Alaskan villages.

———————

Centrally located in the old village, one of various elders with the Noatak Friends Church will ring the huge bell—on a tower next to the river—when services are to be conducted. On the quiet Sunday that we arrive, it's obvious that the Sabbath is widely honored by villagers. Still, with a fuel-oil shortage, the church is so cold that Ricky is challenged to finish his prayers.

The culture of Christianity is pervasive in Noatak. One evening, gathered on a carpeted floor, we laugh with several Iñupiat women—all church-goers—who are engaged in a competitive game of Rummikub (played with tiled numbers that are matched with other players' numbers). Chris and I get roundly trounced. To make sure they win, the women good-naturedly use their fingers to draw signs of the cross on the carpet between themselves and the opponents whom they want to jinx.

Back at the Booths' neighborhood, their next-door neighbors open their windows and crank up the television's volume tuned to a fire-and-brimstone station—the impassioned preacher's exhortations can be heard a hundred yards away. Although the irony might be lost on

Thurston, he shouts over the blare of his TV that he has always loved cowboy and Indian shows. I can't help but think that the noise is also his answer to the neighbor's fervent broadcast. With more than six hundred episodes of *Gunsmoke* and innumerable other cowboy shows on his satellite's Western Channel, Thurston can easily hold off the loud salvos from next door.

For the next couple of days, Chris and I eat every meal with the generous, high-spirited Booths. Dinners feature everything from hamburgers to blackened meat (seal intestines), muktuk (beluga), and the preferred tuktu. Sides include mashed potatoes, French fries, and spinach—since there's a shortage on fuel oil, the Native Store had to shut down its freezers and give away the frozen spinach (which is temporarily substituted for sour dock in the Booth household). The stout Thurston—proudly bare-chested in the mornings like Ricky—serves us huge pancake breakfasts. The Booths insist that we use their shower.

As in many Iñupiat homes, Hilda and Thurston have adopted young children (whose real parents cannot properly care for them) along with grandchildren who happily dash in and out of the house. In lieu of a playground, they cavort on a huge, bowhead whale jawbone out on the riverbank in front of the house.

They moved into their home in 1974, a decade after the diesel generator station first brought electricity into the village. Thurston, 67, believes that "global warming is a big thing for us and should be our first priority." He talks about how the caribou no longer migrate past the village and that the beavers arrived two decades ago. "We're losing the trout [Dolly Varden]. The beavers pollute the water."

Thurston laments from his comfortable seat on the four-wheeler that the salmonberries have dried up and gotten a lot smaller in the heat. "Ten years ago, it changed. You didn't really have to look for them, the berries were all over the place," he points to his backyard. "Now we have to go all the way to Cape Krusenstern to find them." (An all-day boat ride.)

While Noatak is over forty miles upriver from the ocean, the lower jawbone of a bowhead whale that the boys frequently play on shows the Iñupiat ancestral connection to the sea and its animals. Iñupiat ancestral lands. JON WATERMAN

The list goes on.

"Now we can't see the forest from where we're standing because the willows came about ten years ago. We just can't figure out how they got so big."

A half mile downstream, alongside the tangle of willows, the Noatak has cut into and removed more than a hundred feet from the village riverbank and taken away a road. Less than two hundred feet of ground remains between the flood-prone river and the airstrip. Since the 1990s, when the river shifted course and put a stop to the semi-annual supply barge from Kotzebue, airplanes have begun to provide most of the village supplies. Coupled with high fuel costs to fly in fuel and food, Nautaaq's lifeline to the outside world is compromised. The village gas station (a shed with diesel and unleaded fuel tanks) is empty.

Still, the traffic noise from the four-wheelers (who stocked up on fuel) is what you'd expect from a village several times bigger. While quiet in the morning, the four-wheelers—to Ricky's annoyance—zoom back and forth past his plywood cottage at the end of the village until midnight. Like more than a few houses in the old village, a portion of Ricky's house tilts cattywampus on the thawed permafrost ground.

Chris and I take long walks around the village. Now that we're done with our river paddle, the water has turned blue and temperatures have risen. Amid the densest mosquitoes we've seen yet and fist-sized road cobbles that shift underfoot, movement in the village is a hardship without four-wheelers. With fewer than a dozen cars or trucks in a village often swept by icy winds, the four-wheeler is an essential tool.

On the edge of town, the public school is mostly staffed by white schoolteachers who are well integrated into the community. The principal kindly shares the wireless password with us so we can message our families. Like other newer buildings, the school—built in 2008—is jacked up several feet into the air above thick, gravel foundation pads that help alleviate a thaw in the permafrost ground below. The jacks can raise or lower the building as the permafrost thaws or shifts.

Water is piped to newer, plumbed buildings in the village from the water treatment plant (equipped with a small photovoltaic solar system). The plant, built in 1995, had cool-water condensers under the concrete foundation pad to prevent permafrost thaw. Still, less than a decade later, the foundation has begun to sink. Now, with cracks spiderwebbed across the concrete floor and walls, sections of the floor dip as much as half a foot.

Throughout the village, water lines have begun to break as the permafrost thaws and shifts. The pipes from the water plant to Ricky's house broke months ago, so he carries his water home from the Booths'. One more reason he needs to skedaddle back to the wilderness cabin where his stream runs year-round.

Next to the plant, we visit with Vince Onalik, 54, who sits on his boat in a graveled front yard. "Whew," he says, as he leans on his cane, "it's hot. My body is all out of whack."

Like many villagers he wants to talk about climate change. "We're seeing shorter cold snaps, and last winter we only had a month of thirty to fifty below," Vince says. "In the past it was cold the entire winter but now we have rain in December. We had snow in July. That hardly happened before."

He describes how they'd never seen a hundred-degree summer heat before. Even twenty years ago it was rare to see the temperature exceed seventy, he says.

I ask if he has hope.

"Hope?" he replies wearily. "Nothing we can do but trust the Lord. Yet maybe people down south could lower their emissions."

Kotzebue

August 2, 2022

1   Before the 1980s, Kotzebue Sound had been filled with thousands of beluga whales. Now there may be several dozen of the Kotzebue Sound stock of belugas left. Pressure from hunters and an increase in motorboat traffic are the most likely causes, which also explains similarly plummeted populations elsewhere in Alaska, such as in Anchorage's Cook Inlet, where the beluga is listed as endangered. It's uncertain how much climate change has affected the genetically distinct population of Kotzebue Sound whales. Beluga populations in the Chukchi Sea beyond Kotzebue, however, remain healthy.

On a sunny afternoon we catch a ride with Robbie Kirk to Kotzebue on his riverboat. Robbie kindly declines our offer to pay him for fuel or the ride. In demand to haul people and supplies from Kotzebue to Noatak, Robbie—who grew up in the Arctic and lived with his mother for a spell in Texas—is sanguine about the climate crisis. Like Iñupiat everywhere, he faces what could be catastrophic changes with stoicism and grace.

Seals and walrus, he says, are still plentiful along the coast. And later freeze-ups on the river lengthen the time to seine for fish.

Chris and I bounce up and down outside on the chilly bow deck, while Robbie laughs with two Iñupiat passengers in the heated cabin, and we all watch osprey flit above the river. Periodically, Robbie passes around bags of dried salmon and moose jerky. After three-hundred-plus miles of westward flow, the river turns south (before Noatak) and meanders just outside the western boundary of the preserve. The banks are thick with alder, dwarf birch, and for a time, spruce. Twenty miles from the sea, ringed seals shyly pop their heads above the water as they hunt for salmon in the inbound tide.

The ride smooths out in the glassy salt water of Kotzebue Sound and Robbie aims for the scantly visible town, six miles off across open water. I look for but don't see any beluga whales.[1]

PREVIOUS SPREAD: Twenty-six miles above the Arctic Circle, Kotzebue (named after a Russian explorer) is the bustling service and transportation hub of Northwestern Alaska. The town has been occupied by the Iñupiat for thousands of years. JON WATERMAN

Kotzebue lies on a gravel spit of the Baldwin Peninsula, a seventy-five-mile-long finger that points out from the palm of the Seward Peninsula into Kotzebue Sound. The town is the central hub for villages on the Noatak, Kobuk, and Selawik Rivers to the north and east, with another dozen villages to the south and west collectively inhabited by about five thousand people.

Ideally situated for marine mammal harvests, the gravel spit of Kotzebue (population 3,100; called Qikiqtaġruk, or Big Island, in Iñupiaq) has been occupied for thousands of years. Today it has the largest schools in the region, along with an Iñupiaq language school for pre-kindergarten through first grade, but otherwise Iñupiaq is rarely spoken. Sick villagers often boat or fly in from the bush to receive care at the Maniilaq Health Center (Maniilaq is the widely known and revered nineteenth-century Iñupiat prophet from the Kobuk River).

The town is named after the Russian officer Otto von Kotzebue.[2] Long before the Russians owned Alaska, Asian people had crossed the Bering Strait, along with other Iñupiat who traveled up and down the coast, to trade for the region's rich surplus of animal food and furs. This cash-and-barter economy endures in Qikiqtaġruk because the cost of flown-in groceries and supplies stretches those residents not employed in one of the many government services in town. A gallon of milk costs ten dollars and groceries are 35 percent more expensive than in Anchorage (where a gallon of milk costs four dollars).

Subsistence users now contend with thin, honeycombed sea ice that is dangerous to walk on, versus floes thicker than men were tall in colder times. Then there's the decline in the Western Arctic Caribou Herd, and the newly arrived beavers that, a knowledgeable local tells me, have dammed every stream across the peninsula.

In 2015, Barack Obama came to Kotzebue, and became the first president to visit the Alaskan Arctic. Most of his twenty-minute speech

2 Otto von Kotzebue searched the sound for the Northwest Passage in 1818. Near the town later named after him, he examined a bank eroded by the sea to expose a hundred-foot-tall cliff of permafrost with the "purest ice" that held a "quantity of mammoths' teeth and bones." Kotzebue's book *A Voyage of Discovery* is an incredible account of his encounters with savvy Iñupiat along the coast we traveled. The Russians laid claim to Alaska in 1741 and sold it to the United States in 1867.

centered on the climate crisis: "Think about it," he told an audience of hundreds in the high school auditorium, "if another country threatened to wipe out an American town, we'd do everything in our power to protect it. Well, climate change poses the same threat right now.... [W]hat's happening here is America's wake-up call. It should be the world's wake-up call."

None of this came as news to locals. But with imitated seal barks, or their hands, they applauded the president's initiatives to study the climate crisis and create new jobs.

In the evening, we visit Seth Kantner, 58, a white commercial fisherman raised in a sod home 150 miles up the Kobuk River. He comes to Kotzebue for half the year, while his parents moved to Hawai'i to subsistence farm. Seth continues to return to the Kobuk wilderness home that anchors his life.

A storyteller, lithe and short in stature, he pulls out a musk ox skull and horns and self-deprecatingly recants how hard it was to put on a good show for his strong companions as he picked up the heavy piece off the tundra as if it weighed nothing, and then packed it out of the wilderness. His adult daughter, China, looks on with appreciation; Chris and I laugh. Horns cradled in his arms, Seth sits on his bed in the one-room "shack" (as he calls it) that serves as a refuge for sleep and photography work. He repeatedly hands us cans of ale, the first beer we've seen in a month.

"Up the Kobuk," he says, "the animals continue to come to me" to sustain his subsistence lifestyle. But he talks wistfully about the changes: how the tundra landscape that once surrounded the sod home where he grew up is already surrounded by a spruce forest. How the blueberry bushes have been overtaken by Labrador tea shrubs that are three or four times higher than they used to be. "The animals and plants," he says, "are my companions and I've watched them get old."

The novelist and commercial fisherman Seth Kantner—with a musk ox skull and his daughter China behind—spends half the year in Kotzebue between living out on the land in the sod home he grew up in on the Kobuk River. JON WATERMAN

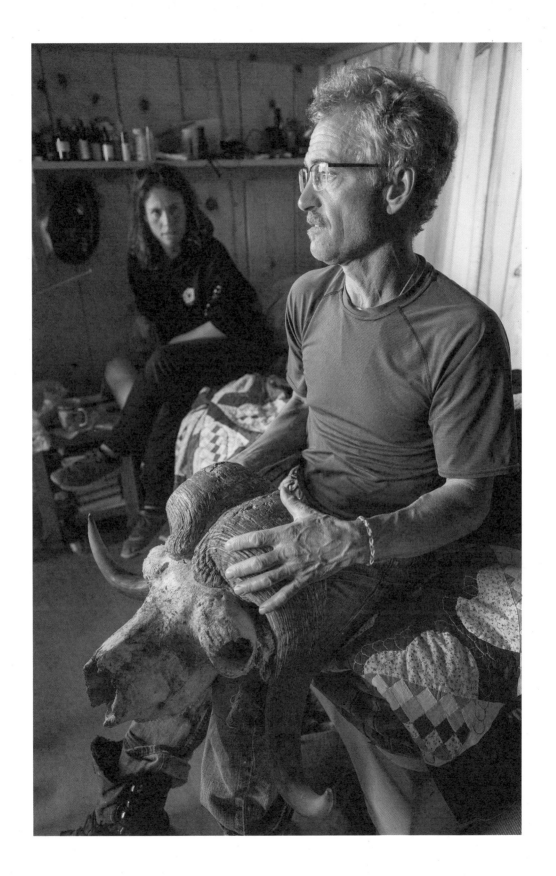

The recent uptick in boat and airplane traffic has made the once-tranquil river noisy. Along with other longtime subsistence users—both Iñupiat and white, whom he calls "Arctic people"—Seth talks about how they no longer recognize landmarks along the overgrown river. "Alders are now everywhere," he says, "big and scary and hard to get through." (Scary because bears often lurk inside these dense thickets.) "What used to be foot-high dwarf birch is now up to ten feet tall."

In between mouthfuls of popcorn, Seth decries the roads that now "chop up the wilderness." This includes the Red Dog Mine Road that we'll pass in a few days, along with proposed roads into Noatak Village, or into Ambler for a huge copper mine. "Roads will only bring more rules and regulations that the Iñupiat culture already despises." Never mind the harm to wildlife and to his subsistence way of life.

Roads aside, Seth used to hunt caribou when they were still fat in August or September and not need a freezer because early falls were cold. Now the meat can spoil in the warmer temperatures. He can scarcely hunt them anymore in the fall because the migration is much later, and the meat is no longer palatable after the caribou rut in October. Their fur that he once bartered or made his clothes or blankets from is also torn up from the new brush.

"The biggest thing," he says, "is trusting the ice, figuring out when it's dangerous. And I can't set nets anymore for white fish under the ice, either.

"It's so much warmer than it used to be. I live in fear of the next cycle where we get a terrifyingly warm winter," he says. "We travel on ice, and we count on snow to be dry not wet. Snow was our insulation. It's important for your house to get buried."

Seth attended the University of Montana and has written several acclaimed books—such as the bestseller *Ordinary Wolves*—a fictional account of his life in the North. He is worried about the future. "I'm not a big believer in science; there's too much arrogance around it. I don't think science is going to save us.

"We're scared of rain here. We are not rain people. That winter seven years ago was an absolute mess, everything was coated in ice, the musk oxen, caribou, and Dall sheep were all cut off from their food." When he effortlessly sets down the musk ox skull, one notices arms made ropy from lifelong labor with heavy fish nets or hefty caribou.

"We start getting winter rains and no ice," Seth says pointedly, "that might be the end of who we are as Arctic people."

Journey's End at Kivalina
August 3–10, 2022

We hitch a ride again with Robbie Kirk fifteen miles north and across the silken waters of Kotzebue Sound in the calm before the forecasted, big-surf storm. On the far shore, we help him lug a generator into his cabin. His Iñupiat companion—a revered elder—is concerned that we don't have a firearm for the grizzlies that we'll meet along the coast.

In their younger days, these men spent months every year out here on the sands of Sisualik (the place of beluga whales). They netted and dried fish on racks, and hunted seals, walrus, and beluga that once used the shallow waters to give birth, molt, and feed on fish. Now the thinned sea ice and change in seasons have begun to upend these age-old traditions at Sisualik. Iñupiat still try to harvest nearly fifty different animals and plants here.

The white canvas tents and skin boats of past decades have been replaced by four-wheelers and scores of locked cabins and plywood shacks. Several lean precariously down a sloped beach, wave battered and on the verge of total destruction as the Chukchi Sea laps ever higher up the shoreline. As the winter ice pack (that quells waves) continues to thin, and shore-fast ice dissipates, storm surf routinely breaks on and erodes the beach, soon to be inundated by sea-level rise.

PREVIOUS SPREAD: Surrounded by the Chukchi Sea and a scant seawall, the village of Kivalina—no longer protected by sea ice and threatened by rising ocean levels—will soon be under water. CHRIS KORBULIC

Thirty thousand years ago the Arctic Ocean did not exist here. In place of waves was a sea of grass grazed by all matter of wildlife on a thousand square miles of land that stretched to Asia. We could have walked for weeks across the Bering Land Bridge.

Ten thousand years or so after the sea first receded, Asian hunters chased the animals across the grassy plain. Most common were small horses, followed by the two thousand-pound steppe bison, caribou, and twelve thousand-pound woolly mammoths (among many other species). Today's wildlife populations in northwest Alaska are only a fraction of the prolific megafauna that roamed and punched trails across the rich grasslands of the spacious Bering Land Bridge.

Because of the extremely dry and clear weather of those times, spring started earlier and summer lasted longer. Compared to the modern-day cloudy weather along the coast, blue-sky Pleistocene days made summers warmer, while the fertile soils and lack of trees created a smorgasbord of grasses and sedges for animals that grew huge—despite the colder continental climate.

This was the climate shift of the Ice Age. A natural process of the Earth under change, versus today's climate crisis manufactured by humankind.

Twenty thousand years ago, the only carbon emitted by human nomads came from their campfires. Like their animal prey, these hunter-gatherers traveled east and west across the land bridge that lies beneath Robbie's hull. The migrants from glacial-free Siberia would have been drawn—maybe even entranced by—the brilliant white glaciers that blanketed the Brooks Range. The first sign of these migrants was found near Kotzebue, dated from twelve thousand years ago through charcoal in their firepits, just before the ocean washed back up over the land bridge.[1]

In the Sisualik area, some five thousand years ago, long after most glaciers to the east had melted, ancestors of the Iñupiat settled on ridges

1 The ancient fire was discovered next to a hot spring in the Bering Land Bridge Preserve on the Seward Peninsula in 2005. Among the charcoal were charred animal bones, which were probably caribou that early Beringia nomads butchered and ate at the site. There were also grooved spearpoints commonly found in Alaska that have not yet been found on the Siberian side of Beringia.

We built a driftwood windbreak and called it a day in punishing icy winds that whipped whitecaps down the coastal lagoon and ancestral Iñupiat hunting lands that we would paddle the next day. CHRIS KORBULIC

above the beach. Even after the megafauna disappeared, the people flourished here throughout the millennia as marine mammal hunters.

Years before President Carter set aside this land and seascape as Cape Krusenstern National Monument in 1978, archaeologists began to uncover caches and old houses of the ancient hunters. For two hundred generations, more than a thousand nomads gathered here and on the Kotzebue spit for their midsummer trade fair. But today (and in the interminable days that follow) as we slog through sand, we will meet only one elderly Iñupiat couple out to gather berries.

Sisualik is separately zoned as private Iñupiat lands within the national monument. As we load our cart on bicycle wheels at 8 p.m. and Robbie's boat disappears in the blue distance toward Kotzebue, we're dismayed at the loose sand. We sink in ankle-deep with every step. Led to believe that the coastal surface was hard as concrete, it turns out that the unusual summer storms have piled up loose sand on beaches seldom sheltered by sea ice. After a mile lashed to the cart like Sisyphus with his boulder, I realize that we'll be in trouble if we have to continue on foot another eighty miles to Kivalina.

Amid thick mosquitoes, we cut and lash strips of plastic flotsam on the bicycle wheels to improve the cart's flotation, but it makes no difference. I find an abandoned kiddy sled that allows me to transfer some weight out of my pack and into the sled as Chris—so much stronger than me—takes over the cart. We continue to stumble along in deep sand at less than a mile-per-hour pace, as seals burble alongside and watch with what I imagine is amusement. Dolphins surface and blow.

Sometime after midnight, in coruscated sunlight, we call a halt. My back and hip are locked up; I lie down on the sand and try to stretch. Chris is also tweaked.

Loons cry forlornly from the lagoon behind the beach. Sandhill cranes rattle, croak, and trumpet their superiority. Chris writes in his journal:

"Brutal making progress." He notes that we struggled fewer than four miles in over four hours.

I lie awake in the tent, dosed with prescription anti-inflammatories. I wonder if I'll be able to walk in the morning—Type 2 Fun has resumed.

In the morning, we continue to plod along grizzly tracks that pockmark the beaches. Above our heads a jaeger attacks a gull and in an aerobatic dance of twirls, swoops, dives, and repeated pecks—accompanied by squawks—the gull drops back into the sea, chastened, to join a raft of gulls that stick to the water. The air, at least for now, is owned by the swift-winged, predatory jaeger.

After little more than a mile of thankless labor in front of cart and sled, we realize that the southwest storm winds will give our hard-to-steer packrafts—not suited for sea travel—a needed boost. So, we load up the rafts, squeeze our bellow bags to inflate them, pull on our PFDs, strap the cumbersome sled and cart atop the boats, push into the sea, and begin to paddle our bulbous crafts with the wind at our backs. And it works!

We paddle all day. A grizzly appears at the washed-up carcass of a beluga and sprints off as soon as it smells us in the wind. A couple of miles north, from inside the rotted pile of a small bowhead whale, another bear emerges—I paddle directly toward it and before I can say hello, it makes for the hills in fright.

I take the flighty grizzlies as good news. "I've never seen bears abandon dead animals before," I say to Chris.

Dark clouds build in the south, and Chris sets the pace as waves rise beneath our boats. As the rain clouds catch us at a wrecked driftwood cabin, I suggest a bivouac, but Chris won't hear of it. I follow, reluctantly, with the knowledge that I'm out of gas: exhausted, sore, and in need of rest.

Chris wheels the heavy cart through beach grasses—away from the soft sand on
the Chukchi Sea beaches, lit by the midnight sun along the coastal lagoons of Cape
Krusenstern National Monument. JON WATERMAN

We continue for another two hours. The surf builds and the sky darkens as we round a point. In the wind and waves, our boats blow in toward shore, so I fight the drift with the bow aimed out to sea and forge an awkward, diagonal progress. I stroke hard against every wave to straighten out and stay upright and imagine that Chris, whom I can no longer keep up with, is unfazed. The waves frequently wash over our spray skirts but, with well-timed paddle strokes, we can keep the tubby boats in motion. Finally, another cabin appears and I put on a burst of speed to catch Chris.

"Shall we stop?" I shout above the wind.

"No," my laconic companion says facetiously with a half-smile. Which means that he, too, has finally peaked.

I pull in first and time my ride with the last wave in a set. It pushes me sideways up the beach and as I rip out of my spray skirt and step out onto the littoral, another wave unexpectedly pushes my feet out from under me and upends me onto my back and into the water—from my knees, I snag the boat as it washes back into the sea. I'm now fueled by adrenaline.

Without a dry suit (which we mailed home from Noatak to save weight), I'm soaked. Chris, of course, surfs in flawlessly, nimbly jumps out, and pulls his boat from the shore break without a splash.

As the rain lashes sideways inside a cold wind, we walk a load of gear up to the locked cabin. Without further ado, on the verge of hypothermia, I strip and pull on dry clothes in the cabin's lee. As Chris goes back to grab the tent and his boat, I look for a key and immediately find it hidden beneath the steps. Within a half hour, we fire up the woodstove and build a line to dry wet clothes.

For two gray days of storm winds and rain, we gather driftwood, read, and fill out sudoku puzzles. As I scoop up buckets of melted permafrost water—browned with plant tannin—I find much-needed wonder in brilliant yellow poppies that shiver in the breeze. And in a colony of terns on the beach that pump their wings and fly about my head with shrill, rachet notes.

Chris writes in his journal: "I know I'm approaching the end of my trip when I start planning chores at home." Sandhill cranes jump up and down outside on the tundra. The cabin shakes in the wind as we sleep on caribou-skin blankets.

———————————

The prophet Maniilaq was born along the Kobuk River in the early 1800s. After a bird spoke to him through a song that was the "source of intelligence," his mother told him that he would become a shaman. But since his power came from a different place than the mean-spirited shamans, Maniilaq rejected their hold over the people.

Like later missionaries, he went to Sisualik and showed the hunters how to break the taboos and eat muktuk mixed with land foods. While the people fearfully moved away from him, he didn't become sick or die.

Maniilaq called the creator Grandfather and began to foretell changes that would take place. He said that people would come from the south and east with white-colored skin. And that they would bring books for the people to understand.

Maniilaq traveled extensively and studied the land and the animals. He raised a family. Before he disappeared somewhere in the east, he warned the people. His prophesies spoke of crafts powered by fire that would go up the rivers and carry people through the air. One day, he predicted, in the place near Ambler, the village would become a great city and people would pull things from the ground that would make them rich. He predicted a famine.

As told today by many elders in Kotzebue and other villages, he prophesied that a time would come when a season would occur twice in succession. The prophecy was vague and could have meant that two summers or two winters would merge together. Whatever the case, many Iñupiat believe that Maniilaq had been a flesh and blood man, as real as Jesus. They believe that Maniilaq had, in fact, foreseen the time of starvation caused by whalers, the time of the missionaries with their Bible books, the mines, and the climate crisis.

—————

We walk all day toward ore ships the size of city blocks at anchor a half dozen miles away from the Red Dog Mine port. At the bigger streams we blow up our boats to cross, but without a south wind, it would be a jester's errand to paddle our river rafts in the ocean. We see a few more dead gulls on the littoral, but nothing like the die-offs reported from previous hot summers (we may have missed carcasses on the beach when we paddled in the lagoons and out at sea).[2]

It takes us all day to plod a dozen miles. We stop, wearily, in the lee of a wrecked cabin frequently inhabited by storm waves, but now home only to a few swallows. A sofa and beds, along with tools and clothing, are abandoned inside walls held up by washed-in sand. Mosquitoes buzz about in the cold.

A few miles north, we can see the massive dock connected to the Red Dog Mine road. Beyond is a warehouse—garishly painted red, white, and blue—that could fit a hundred houses. Trucks rumble back and forth. Combined with the ships offshore, the scene of modern industry is as unreal and out of place as if we'd just stumbled down the yellow brick road to confront Oz. This is the port and access to one of the largest zinc mines in the world.

In the morning, we stagger into a mean headwind before we finally cross the mine road.[3] Just like my experiences at Prudhoe Bay—emerged from the wilderness and out alongside hard-hatted men at work—a dozen Asian men debarked from a ship keep their eyes straight ahead as if we're invisible, without acknowledgment of our presence.

[2] Before 2015, it was rare to find seabird die-offs in the Chukchi Sea. But since 2017, per recent USGS and NOAA reports, Natives from Point Hope to the Aleutian Islands have repeatedly observed large numbers of dead auklets, shearwaters, murres, puffins, kittiwakes, and various ducks in the summer. In 2022, over four hundred dead birds—a quarter of the numbers reported in 2017—were found along this stretch of coast. They were presumed to have died from starvation. In addition to public health concerns for subsistence communities, the scientists wrote of "a harbinger of concern for the state of the Arctic Ocean itself." No specific cause for the starvations was listed, but no one disputes the significantly warmer temperatures and loss of sea ice. The oceanographer Dr. Peter Winsor, one of several proofreaders for this book, suggested that the starvations were "likely due to changing stratification and food sources being forced deeper down in the water column, out of reach for the seabirds."

One of many dead birds—a herring gull—that have begun to wash up on Chukchi Sea coasts in recent years as the ocean temperatures have warmed and caused starvation.
CHRIS KORBULIC

3 The road leads fifty-three miles (through twenty miles of national monument lands) to a huge open-pit mine and settlement ponds. Red Dog Mine sits on land owned by NANA, the Native corporation owned by local Iñupiat. NANA had been granted the land through the 1971 ANCSA legislation. After the corporation agreed to the mine, it shares profits from mineral sales, along with jobs for Iñupiat shareholders. But pollution and interference with wildlife make for an uneasy alliance between local Iñupiat and the mine.

The June 2022 issue of the *National Library of Medicine* journal reported heavy-metal dust (zinc, lead, and cadmium) out to a third of a mile on either side of the road on park service land that impacted plants such as the lichen that caribou depend upon. Then a decade-long study published in a 2016 issue of *Biological Conservation* showed that about a quarter of the Western Arctic Herd stopped at the road in the fall and delayed their southern migration.

Maybe they're worried we'll beg them for food or water or rescue, I think. On the opposite side of the road and titanic dock, we stop and brew coffee inside a twenty-foot container from a cargo ship, provided for passersby. Aside from the occasional Iñupiat hunter, it's hard to imagine the shelter sees any traffic.

Annoyed by the modern industry on these once-wild shores, and plagued to make miles under difficult conditions, there's scant opportunity to enjoy the beauty that exists here along this remote Arctic coastline. Chris is obviously done with the trip and just wants to reach Kivalina. Conversations are truncated. I don't take it personally and abstain from conversation that annoys him.

We leave the container reluctant to head back into the wind. Chris charges off like a racehorse in front of his buggy, and we squint against plumes of sand blown aloft. Forced to stop at another unfordable steam, we pause for the night a dozen miles south of Kivalina.

Chris's hands are puffed up from mosquito bites and saltwater immersion. He, too, has limped all day, plagued with back pain. As the wind hammers our tent, he writes about our departure from the weird mine area: "Continued to two miles of leg and soul punishing beach to windy camp at end of ten-mile lagoon. Built driftwood wind block, took 1000mg ibuprofen + muscle relaxer and finally slept great with little pain."[4]

The next day, as Chris wordlessly blows up his packraft and begins to paddle the lagoon, I make a quick plea that we discuss the day's strategy as we go along. "We paddle," is all he says, then smiles and takes off. It occurs to me that my companion might just be in constant pain, which would explain the reticence.

As I try to hold the pace, I consider the dilemma of expeditions—whether up mountains, across seas, or down rivers. There is always the urge to find your limits, to push as hard as you can, and then somewhere along the way, you aspire to smell the wildflowers. Granted, our itinerary has

4 Chris had rolled on a ball most of the trip with the assumption that his pain had been caused by irritated IT bands. But when he got home, he learned that he had ruptured discs in his back. Shortly afterward, he underwent discectomy surgery to remove the bulged portions of discs that protruded from his vertebrae.

When I got home, X-rays revealed that I had compression fractures of five vertebrae, undoubtedly exacerbated or caused by the long paddle days and the deep-sand sled haul. Further X-rays showed that I also needed two knee replacements and surgery for severe arthritis in my thumb.

been hopelessly ambitious—to cover so many miles and simultaneously document the climate crisis—but as a graybeard behind a young man at the height of his powers, what choice do I have but to ride toward Type 3 Fun? If I had chosen most any other partner, the trip, amid such unseasonable cold, would've ended at Noatak Village.

That evening as the wind dies down a few miles shy of Kivalina, I convince Chris to stop and relax. With cameras in hand, we comb the beaches under a meek and reluctant sun. I watch a soon-to-perish bumblebee attend business amid a colorful, yet almost-wilted, last-gasp splurge of coneflowers, Shasta daisies, and sunflowers. Dwarf fireweed along the lagoon shores has now flowered to the top of its otherwise bare purple stalks to signal summer's end.

A stone's throw away on the Chukchi Sea beach, fresh grizzly prints enliven silvery, crystalline sand next to tufts of grass that have etched perfect circles around themselves—as if protractors had drawn lines in the sand—to show the last few days of wind. In the waist-high sedges behind the beach, we find yet another whale bone, dense and heavy as iron.

Announced by the whine of their engines, three Iñupiat couples pull up on their evening ride out from Kivalina—thanks to a newly built evacuation road off their island—and shut down their four-wheelers. All six are dressed in winter jackets, wool caps, and gloves. Curious to find us camped in their backyard, they offer us coffee from thermoses, and we trade stories. Reppi and Dolly Swan say that they haven't yet seen summer, even though it's now August 9. They invite us to come visit them in town.

In the morning, I realize that, more than any place I've visited, Kivalina embodies the Arctic climate crisis. The first thing you notice about the barrier island—surrounded by a wide lagoon and the surf-driven Chukchi Sea—is its half-mile-long, six-foot-high seawall. The rock wall stands as the village's only protection against the sea, which the Army

Over the millennia, in coastal sites like Kivalina, Iñupiat gathered and grasped sewn-together seal skins—the world's first trampolines—to bounce a villager high into the air to scout for whales out in the sea. JON WATERMAN

Corps of Engineers predicted would inundate Kivalina by 2025. The revetment seems tenuous protection for an ocean now ice-free for much of the year.

In 2006, after the Army Corps suggested that the village be moved away from the sea, Kivalina filed a lawsuit against nearly two dozen energy conglomerates (that included ExxonMobil, BP America, Chevron, and ConocoPhillips). Since villagers believed that Big Oil created the climate crisis and withheld information about it, they asked for $400 million to pay for the cost of their relocation off the island. Dismissed by a district judge, the village appealed to a higher court, but by 2013 the case was similarly rejected because the villagers lacked the evidence to link their injuries to the companies' actions.

Eight miles long, the island is only about seven hundred feet wide and tops out at ten feet above sea level. To the east, the half-mile-wide, deepwater lagoon has eaten away at the land alongside the houses and as we paddle in, the shoreline is slumped into wavy hummocks from permafrost thaw. Since there's no hotel, the village manager puts us up in the community center for a couple of nights at a price that beats any motel in Anchorage.

Children run all about the tiny village, where they play catch, tumble about on landlocked boats, and bounce on trampolines in a modern-day reenactment of the traditional Iñupiat blanket toss. In midair the teenage bouncers read texts from smartphones clutched in their hands.

There is no plumbed water or septic system in Kivalina, so honey buckets are emptied on the far side of the island. Since the village lies sixty-six miles downstream from the Red Dog Mine, few trust the water taken from the Wulik River,[5] so the community center is stocked with jerricans of water melted from snow.

A 2019 statewide assessment identified Kivalina as one of seventy-three Alaskan Native villages (out of more than two hundred) to face significant environmental threats from the climate crisis. In a 2003

5 Six Kivalina residents who believed that their fish and water supply had been polluted sued the Red Dog Mine. In 2006, a US District Court judge ruled that the mine had violated the Clean Water Act more than six hundred times for pollution discharged into Red Dog Creek (upstream of Wulik River). Eventually, the Canadian company that owns the mine paid $8 million in damages. Although the mine may have resolved hazardous pollution issues, villagers still complain about excess sediments in the river. To make matters worse, streams in northwestern Alaska— including the Wulik River—have begun to turn bright orange. A story in the January 2024 issue of *Scientific American* hypothesized that the thaw is releasing iron from the permafrost, literally rusting the rivers. (On our journey, we only saw a few isolated orange puddles and small colored streams.)

report issued by the Government Accountability Office in DC, Kivalina was listed as one of four villages in imminent danger, which prompted the construction of a wall consisting of rocks in wire-mesh baskets in 2006. Two hours before US Senator Ted Stevens arrived to cut a ribbon and celebrate its completion, the wall had already begun to wash away. Before it disappeared, the faulty wall only increased shoreline erosion.

"The beach used to be quite a distance away," Reppi tells me. "We used to pile everything we could on the beach. One time a plane crashed, and we tore it up with an excavator and threw it out there."

The Army Corps of Engineers finished the present-day wall in 2010 with the idea that this would buy the village enough time to relocate. By 2021, a bridge was completed across the lagoon and Reppi, hired as a truck driver, helped construct the emergency evacuation road that leads eight miles up to a new school.

Reppi, 48, is a respected hunter and fire chief in a densely packed village that could be entirely burned down from the spread of a single house fire. But these days, locals are more worried about floods.

Reppi is not the only villager skeptical that the four hundred-plus residents here will be able to move up on the hill. Although the school will open in the fall and move the children away from the old school on the island, Reppi doesn't believe that people could ever live on such a wind-blasted, snowy hill. The most recent estimate for village relocation is $400 million—the same amount the failed lawsuit sought. But neither the state nor federal government has allocated enough funds for such a massive project. (Over the last century, the people have repeatedly moved the village around the island.)[6]

We eat and drink tea with the Swans. As a television blares in the background, Dolly asks about my children. Given that the village has been inundated with visits from innumerable journalists and most major, disaster-hungry news channels over the last two decades to chronicle the expected inundation, Dolly and Reppi—along with other villagers

6   In 1905, the federal government constructed a school on the barrier island because of its accessibility by water. Like Noatak and many other villages in the early twentieth century, nomadic Iñupiat came down out of the hills or more protected coastal locations to put their children in school. Prior to this, Iñupiat only used the island as a seasonal camp. In 1911, a schoolteacher wrote in his official report that the new village "is very beautifully situated when the weather is nice and calm, but when the wind blows from the south it raises the water in the ocean until it sometimes almost comes over the banks. ... We believe that to move would be the wiser if not the safer plan."

we meet—are patient, generous hosts. We're probably the only writer-photographer team that has arrived overland.

As usual, I ask a lot of questions. Chris, quiet as a lemming, shoots videos and pictures.

Crosses and likenesses of Jesus adorn the walls. Polite and deferential to the adults, their six children crowd in at the dinner table with us for fried chicken and muktuk. Reppi raves about how hot it got last summer. "A hundred degrees!" he says. "A new record. We bought an air conditioner."

In 2007, along with other Kivalina residents, the Swan family witnessed their first thunder and lightning storm. It reduced their two-year-old daughter, Rogina, to tears.

Somehow, inside less than eight hundred square feet, they manage to share their house with six children. Their oldest, Sakkan, drives a fifty-foot-high ore truck at the Red Dog Mine—as Reppi did several years earlier. He figures that a half dozen villagers now work at the mine (earning six-figure salaries) to support their families.

Reppi's grandfather was a reindeer herder in Kivalina. By 1950 the reindeer experiment failed across northern Alaska as the domesticated animals went wild and joined the Western Arctic Herd. Social welfare partially filled in food-security gaps that existed from their semi-subsistence lifestyle.

To woo his mother, his father chased and caught a wolf, which is all it took to convince her parents to let him marry her. "Those were the days," Reppi says, with smile lines that crinkle alongside his eyes, "where they still used skin boats and dogsleds."

"With cli-*mate* change it's real hard to figure out how the weather is go-*ing* to be," Reppi says with the singsong phonetics and inflected end vowels that I've come to revere over the last few decades of northern travel. "All the know-*ledge* my father taught me about the weather is now out the door. It changes like that," he snaps his fingers. "It's good then it's suddenly storm-*ey*."

"We had rain in Januar-*ee*!" Dolly says.

After dinner, we board Reppi's white Toyota pickup truck, which after it was transported here by barge, makes it one of the most expensive small pickups on the continent. It's one of six cars or trucks in a town ruled by four-wheelers. He drives over the bridge and at no more than twenty miles per hour up the sharp-rocked, frequently culverted evacuation road. A dozen feet below, the soggy tundra is splotched yellow by salmonberries. There is no one else on the road.

We stop and get out. As Chris photographs the distant sunlit sea, I walk across spongy sphagnum moss, kneel, and cup my hands to drink from the ice-cream headache water of a small stream. I stand up. Then I breathe deeply and let the Arctic open up my heart and fire my imagination like no place on Earth.

Sandhill cranes have begun to flock and chatter and fly in circles to prepare for their migration south. It's time, too, for us to leave.

Dolly and Reppi banter with each other about the willows. "Look a lot bigger this year," he says. "Wish we had trees; they'd be even higher."

"We don't need trees!" Dolly says.

A few miles up we get out again in a cold wind to take in the view of the packed-tight village houses soon to be inundated by the Arctic Ocean. You can't help but feel reduced by the animate enormity of the land- and seascape, which might also explain the gentle forbearance of these northern people. Throughout the millennia, generations of Iñupiat (and Inuit) have had no choice but to adapt to the ebb and flow of ice ages, periods of hunger and deprivation, and shifts of wildlife and weather, and to integrate Christian culture and values into their age-old beliefs. Adaptability and acceptance are practically encoded in their genes. In many ways, the people of the North, with their durable culture, have been the greatest wonder of our trip.

Reppi's village is being defeated by Big Oil and the consumptive emissions of the outside world, while he deals with runaway inflation, forced to take construction jobs to support his family. As a child, white schoolteachers rapped him with a ruler when he spoke Iñupiaq. But Reppi—like most Iñupiat we've met—is not resentful. He has a well-developed sense of humor, free of cynicism or animosity. Nor has he forgotten his father's stories of the old ways.

Reppi, of course, knows about Maniilaq's centuries-old predictions. As recorded by the missionaries, Maniilaq had said that "everything will change." Among the technological and cultural shifts, the Iñupiat seer prophesized that a village surrounded by the ocean would be destroyed in a storm. Still, the people of the North are the toughest people to have ever lived on Earth.

Like our Kivalina friends and other modern-day Iñupiat, Maniilaq calmly accepted the large-scale, long-term dynamics of change and the new crisis that would sweep the land. You can only admire their resilience.

Appendix
Climate Crisis Predictions / How to Take Action

At first take, the Arctic climate crisis may appear as a distant, unrelated phenomenon compared to the changes experienced in the warmer southern world of trees. Yet the transformation of the Arctic has directly affected the southern latitudes through the thaw of sea ice from warming waters, through shifts in ocean currents, and through sea-level rise as the Greenland ice cap melts—among other radical changes mentioned elsewhere in this book. The Arctic is our canary in the coal mine.

A bit of history: By the 1980s the informed world began to confront the irrefutable evidence of human-caused climate change. In 1988, the United Nations and the World Meteorological Organization established the Intergovernmental Panel on Climate Change (IPCC) to assess the damages, present computer-modeled predictions, and make recommendations. Although there are many climate-related organizations that can be consulted, the IPCC is the most authoritative and nonpoliticized source.

In March 2023, the IPCC released a nearly eight thousand-page report with an update on the crisis and what the world can do to reduce greenhouse gas (GHG) emissions that have heated the planet. For this *Sixth Assessment Report* (which can be easily found on the internet) more than seven hundred objective and policy-neutral scientists—who represent 195 nations—weighed in.

The news is not great. The IPCC report states that's it's "likely" that global temperatures will reach or exceed 2.7 degrees Fahrenheit (1.5° C) above preindustrial levels—1.5 degrees Celsius was the threshold set by the internationally recognized 2015 Paris Agreement. As of 2023, Earth's average global temperature has increased by at least 1.1 degrees Celsius since 1880. With another 0.4 degree Celsius rise, the IPCC predicts that 70 to 90 percent of coral reefs will die, along with an uptick in extreme weather events, more frequent and intense wildfires, and more excessive flood and drought cycles. All combined, these climatic changes could lead to a collapse in the world's food systems.

In the Arctic and the Antarctic, temperatures have already increased at least three times more than the rest of the world. This is mostly due to the loss of extensive snow and ice cover that, in turn, allows ground and water surfaces to absorb heat, rather than reflect it.

The IPCC report states that it is "virtually certain that surface warming in the Arctic will continue to be more pronounced" than in the rest of the world over the next seventy years. Fire seasons are expected to lengthen, with sea-level rise "virtually certain to continue" and cause floods and shoreline erosion in villages like Noatak and Kivalina. The IPCC predicts that sometime before 2050 the Arctic Ocean will lose most of its summer sea ice.

GLOBAL ACTIONS NEEDED

Despite dire changes and predictions, there is still hope and a myriad of ways to minimize the crisis. The ultimate goal is to strive for "net zero" GHG emissions—net zero is the state where the amount of greenhouse gas released into Earth's atmosphere is balanced by the amount removed. As called for in the Paris Agreement, worldwide GHG emissions must be reduced to 45 percent by 2030 and as close to net zero as possible by midcentury (with no more than a two degree Celsius rise by century's end).

The IPCC report makes many recommendations that can be enacted on both a personal level and through governmental reforms (along with suggested adaptations). Most of the science-based recommendations hinge upon the elimination or reduction of fossil fuels—the number one source of GHG emissions. Still, even as these words are written, the latest IPCC report is old news (the reports are updated and released every seven to eight years).

On a global scale, the IPCC recommends these steps:

Coal plants must be retired. The world's transportation system needs to shift to mass transit vehicles (or bicycles), eventually powered by nonfossil fuels. Wildlands must be protected, with a halt on deforestation (since trees and other plants absorb carbon that becomes greenhouse gases). Food loss and waste must be reduced, and the world needs to transition to a plant-based diet (see "Food" below). Carbon dioxide emissions must be eliminated from aviation, shipping, buildings, and other essential products (steel, cement, and plastic).

Toward those ends, dear reader, as creatures of habit, it's difficult to change lifelong routines, let alone tell others to do the same. Yet for the sake of future generations, here are suggestions and actions—based on United Nations recommendations and concerned friends from the trip—that will help make a difference. With imagination and thoughtfulness, the list can be expanded.

How to Take Action

DEFEND DEMOCRACY

Like Kotzebue's Willie Hensley (see chapter 7)—who fought for his peoples' rights through the ANCSA legislation—we can urge government officials (through emails or visits to their offices) to shift to renewable energy instead of fossil fuels. Ask local (county, town, or city) officials to set net zero targets for public buildings, operations, and transportation. Share your thoughts and get others to act.

Vote for only those local and national officials who will push for climate action. Take the time to study voter guides and sample ballots. Those candidates who support the fossil-fuel industry, or don't mention the importance of climate initiatives, are only part of the problem. It's imperative to defend democracy—because corrupt or corporate-minded (versus environmentally aware) government leaders will continue to defeat climate change initiatives.

START DIALOGUES

Talk to your family, friends, colleagues, and neighbors; don't hesitate to ask or partner with local business owners to show how they can make a difference (and

PREVIOUS SPREAD: Kivalina—doomed, like many Iñupiat villages—is surrounded by the Chukchi Sea and the lagoon fed by the polluted Kivalina and Wulik Rivers. The recently built road provides an evacuation route to dry ground. CHRIS KORBULIC

295

increase their business as they effect positive change). And when progress is stymied, follow climate activist Greta Thunberg's example and protest, speak louder, and be persistent in order to be heard (please see her 2023 work, *The Climate Book*).

The IPCC scientist Gary Kofinas (see chapters 6½, 7, 7½) believes it is essential "to find ways to constructively engage in dialogue with people who don't want to act or don't think the climate crisis is real. Today we avoid conversations with people who hold different perspectives. It's critical that we develop skills on how to listen, understand others' perspectives, and with mutual respect, find common-ground solutions. The polarization of our social world and inability to converse is killing us!"

FOOD

Although no one likes to be told what to eat, let's face it, a meat diet is detrimental to the planet. As per the National Institutes of Health, "Livestock production does not only have a negative influence on GHG emissions, but also on the water footprint, water pollution, and water scarcity." If everyone switched to a plant-based diet, we could save 76 percent of the planet's land area. Plant-based food production ultimately reduces greenhouse gas emissions. And with the recent development of cell-cultivated meat (cleared for sale by the USDA and declared safe for human consumption by the FDA), meat eaters now have an alternative that is much safer for the environment.

Support local farmers who sustainably farm and grow regenerative organic food—this practice cuts back on the energy used to transport food around the world to

Exhausted, with back injuries, we walk the last miles to Kivalina in an icy wind between the Chukchi Sea and a coastal lagoon.
CHRIS KORBULIC

local grocery stores. And if we cut out food waste and compost our leftovers, our personal carbon footprint can be reduced by up to three hundred kilograms of CO_2 per year.

AT HOME

With the thoughts of Noatak Village's Vince Onalik (see chapter 9), we're overdue to stop fossil-fuel emissions and switch to solar, wind, geothermal, nuclear, or efficient electric energy appliances. Replace oil or gas furnaces with electric heat pumps. Unplug clothes driers and hang clothing (it'll last longer). Use cold-water wash cycles. Switch to all LED light bulbs. Get rid of gas-powered lawn mowers or leaf blowers and replace the water-consumptive, methane emissions of dead grass with a xeriscaped yard.

To get psyched, it's helpful to calculate your personal carbon footprint (there are many useful online sites with footprint calculators). Then as you make changes, you can estimate how many kilograms of CO_2 you'll save a year.

CONSUMERISM

Although no conscientious consumer needs to be reminded to reduce, reuse, repair, and recycle, it's still helpful to reconsider how our purchases of new clothing, electronics, and all manner of newly created goods cause carbon emissions—from the extraction of raw materials to their manufacture and transport to market. One solution: shop secondhand.

As consumers we can also have a huge effect on global emissions through our wallets. Invest in and only use funds, banks, or credit cards that practice environmental sustainability (rather than the many Big Oil or corporate entities that support fossil fuels).

TRANSPORTATION

Electric cars are a passable and partial solution, at least until modern technology can make safe hydrogen cars (I disagree with my wise friend Seth Kantner—see chapter 9½—because this is one of many examples of how science can help save us). Prices will continue to drop and make electric cars affordable, as electric charging stations multiply and begin to work off clean, renewable (instead of fossil fuel) energy.

If you live in the city, carpools, public transportation, or bikes are the way to go. For longer distances, consider trains, buses, or other mass-transit options.

Cut out or minimize airline travel. Emissions from most high-altitude jets stay in the atmosphere with incredibly potent climate impacts that only further heat the planet. If you must fly, conscientious travelers purchase carbon offsets (another temporary solution until air travel can be decarbonized; use only "gold standard" certification).

FINAL THOUGHTS

Chris Korbulic—who lives as a minimalist consumer on Washington's Olympic Peninsula—commented:

"As we heard [on the trip] multiple times, the injustice of industrial climate change is that some of the people most affected, who have contributed least to the problem, have to live with and adapt to the results of many other

people's actions. We all live in a big bubble of the Earth's atmosphere, and what an economy on the other side of the planet does to grow and consume and exploit and emit, we all have to deal with. Will we in the 'climate-conscious' bubble of North America change world economies enough to slow, stop, or reverse this climactic process that is well underway?

"What I like to consider is that there are many, more comprehensible issues close to home that affect local life, and probably also have an effect on global climate. They often seem to be overlooked for the lofty, ambiguous goals of stopping climate change. Where I live, the loss of forest cover is visible, more comprehensible, and most importantly something people here can directly influence at the state and local level. I see rivers being restored, fish populations rebounding, and small bits of land being conserved. I try to live the best I can in a place I love, act to support meaningful local efforts, and contribute within my means to support broader conservation goals."

Nonetheless, the crisis will continue to escalate through floods, wildfires, extreme weather events, and eventual food shortages—all predicted by the learned scientists of the IPCC. In October 2023, the *Fifth National Climate Assessment* report (congressionally mandated every five years) focused specifically on the crisis in North America. Like the IPCC findings, the report—compiled by 750 scientists—warned that, despite greenhouse gas emissions in the United States slowly decreasing, it's not happening fast enough to meet the nation's targets. Nor does it align with the UN-sanctioned goal to limit global warming to 1.5 degrees Celsius.

From an economic standpoint, the climate crisis is already disastrous. The report revealed that US extreme weather events in 2023 (from only January through September) cost $25 billion—more than any other twelve-month period in the nation's history.

There are other huge costs, too. "From Alaska to low-lying Pacific atolls," the report reads, "forced migrations and displacements driven by climate change disrupt social networks, decrease housing security, and exacerbate grief, anxiety, and negative mental health outcomes. Indigenous Peoples, who have long faced land dispossession due to settler colonialism, are again being confronted with displacement and loss of traditional resources and practices."

Indigenous people know that in prosperous states or nations most people will find relative safety and tools of adaptation. This, however, is not true for the world's poor and Indigenous populations with the lowest carbon footprints, who have taken better care of the natural world. In the end it is the Arctic people—along with the disenfranchised in still-to-develop countries, within tropical climates, or at sea level—who will not find shelter amid the storm to come. We owe it to them and to our children's children to act now.

 Scan the QR code for more information about Jon Waterman's trips to Alaska including reader's and teacher's guides, interactive maps and video showing the impact the climate crisis is having on the Indigenous people and the environment in Alaska, and links to the audio- and e-books.

Acknowledgments

Thank you:

Chris Korbulic
Dr. Gary Kofinas
Peter Metcalf
David Schipper
Dave Buchanan
Carolyn Murray
Laura Waterman
Dr. Peter Winsor
Dave Shapiro
Lighthawk
Christine Steel
Katie Homes
Danika Van Lieshout
Dr. Andrew Tremayne
Ricky Ashby
Thurston and Hilda Booth
Vince Onalik
Robbie Kirk
Reppi and Dolly Swan
Millie Hawley
Jeremy Millard
Emily Schwing
Brad and Mary Reeve
Seth Kantner
China Kantner
Doug Pope

Jeff Rogers
Rich Henke
Torre Jorgenson
Dr. Kyle Jolie
Dr. David Swanson
Michael Koskey
Thomas Hassler
Michael Grohmann
Nathaniel Gilbert
Jason Olden
Backpacker's Pantry
Backcountry.com
Alpacka Rafts
Eddie Bauer
Bonfire Coffee
Dr. Roman Dial
Charlie Ebbers
Karla Olson
Heidi Volpe
John Dutton
Elizabeth Kaplan
Christina Speed
Sonia Moore
Bernardo Salce
John Burcham
Stephanie Ridge
Patagonia

A dried-up, thermokarst sluice gully amid a large thermokarst in the Noatak National Preserve has begun to revegetate with grasses and shrubs, which shows the Earth's capacity to regenerate if/when the climate crisis can be controlled. JON WATERMAN

Index

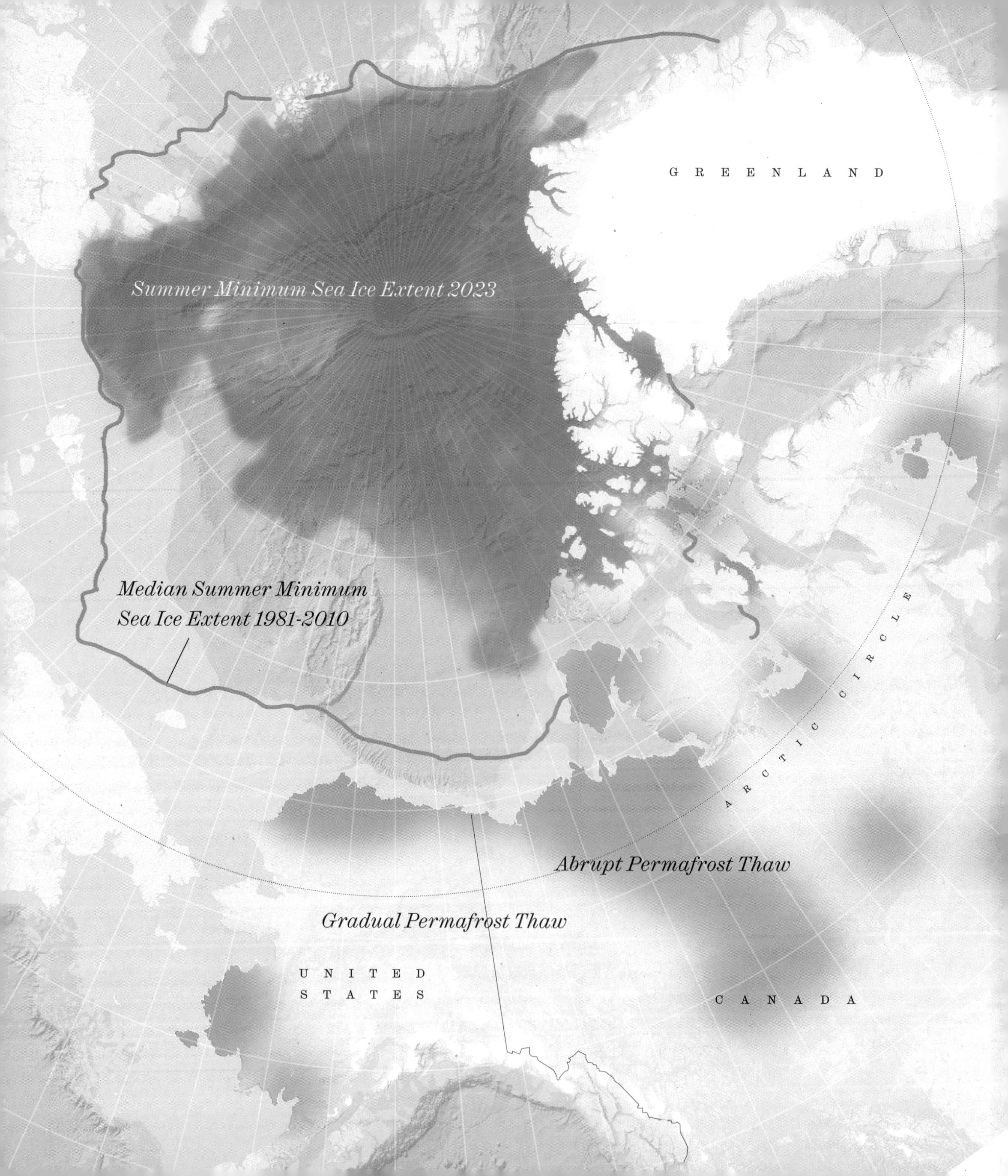

Summer Minimum Sea Ice Extent 2023

Median Summer Minimum
Sea Ice Extent 1981-2010

Abrupt Permafrost Thaw

Gradual Permafrost Thaw

GREENLAND

ARCTIC CIRCLE

UNITED
STATES

CANADA